T0214032

SpringerBriefs in Electrical and Computer Engineering

Series Editors

Woon-Seng Gan, School of Electrical and Electronic Engineering, Nanyang Technological University, Singapore, Singapore

C.-C. Jay Kuo, University of Southern California, Los Angeles, CA, USA

Thomas Fang Zheng, Research Institute of Information Technology, Tsinghua University, Beijing, China

Mauro Barni, Department of Information Engineering and Mathematics, University of Siena, Siena, Italy

SpringerBriefs present concise summaries of cutting-edge research and practical applications across a wide spectrum of fields. Featuring compact volumes of 50 to 125 pages, the series covers a range of content from professional to academic. Typical topics might include: timely report of state-of-the art analytical techniques, a bridge between new research results, as published in journal articles, and a contextual literature review, a snapshot of a hot or emerging topic, an in-depth case study or clinical example and a presentation of core concepts that students must understand in order to make independent contributions.

More information about this series at http://www.springer.com/series/10059

Christoph Guger · Brendan Z. Allison ·
Aysegul Gunduz

Editors

Brain-Computer Interface Research

A State-of-the-Art Summary 10

 Springer

Editors
Christoph Guger
g.tec medical engineering GmbH
Schiedlberg, Oberösterreich, Austria

Brendan Z. Allison
Department of Cognitive Science
University of California San Diego
San Diego, CA, USA

Aysegul Gunduz
Department of Biomedical Engineering
University of Florida
Gainesville, FL, USA

ISSN 2191-8112   ISSN 2191-8120   (electronic)
SpringerBriefs in Electrical and Computer Engineering
ISBN 978-3-030-79286-2   ISBN 978-3-030-79287-9   (eBook)
https://doi.org/10.1007/978-3-030-79287-9

This Springer imprint is published by the registered company Springer Nature Switzerland AG
The registered company address is: Gewerbestrasse 11, 6330 Cham, Switzerland

Contents

Brain-Computer Interface Research: A State-of-the-Art Summary 10

Christoph Guger, Brendan Z. Allison, and Aysegul Gunduz

Abstract Brain-computer interfaces (BCIs) enable users to send messages or commands directly through brain activity, without any movement. Most BCI systems aim to help persons with serious movement disabilities, but BCIs for consumer applications are increasingly prevalent. Each year since 2010, teams submitted their BCI projects to the BCI Research Awards, where a jury decides the best projects of the year. We invite the nominees and winners of these awards to contribute chapters to our annual book series, and this book covers the 2020 BCI Research Awards—the tenth year of this award series, and the tenth book about the awards. Most of this book consists of chapters in which the authors summarize these BCI projects or contribute interviews about their work. In this chapter, we introduce BCIs, describe the procedures and people involved in the awards and books, and present the twelve projects that the jury nominated for an award.

Keywords Brain-computer interface · EEG · ECoG · BCI research awards · BCI foundation

1 Introduction

Brain-computer interface (BCI) systems are gaining attention, with new applications to benefit patients and even healthy users. We started the Annual BCI Research Award in 2010, and have organized this award since then. These awards have had many beneficial outcomes, including: (1) identifying and rewarding the best BCI projects each year; (2) providing positive publicity for top projects and the overall

C. Guger (✉)
G.Tec Medical Engineering GmbH, Schiedlberg, Austria
e-mail: guger@gtec.at

B. Z. Allison
Cognitive Science Department, University of California San Diego, San Diego, USA

A. Gunduz
Biomedical Engineering, University of Florida, Gainesville, USA

C. Guger et al. (eds.), *Brain-Computer Interface Research*,
SpringerBriefs in Electrical and Computer Engineering,
https://doi.org/10.1007/978-3-030-79287-9_1

field; (3) creating an annual book series with summaries of the top BCI projects and discussions of BCI trends; and (4) helping to "raise the bar" for high-quality BCI projects.

This year marks the tenth anniversary of the BCI Research Award and its corresponding books. This book, like earlier books in this series, presents articles from people behind this year's top BCI projects. Most of the chapters were written by these BCI experts and present details about their BCI projects, including: the current state-of-the art; why their work is important; hardware, software, procedures, algorithms, and other methods; results, usually based on real-world testing with patients or other target users; and discussion, often with future directions and other commentary. This introductory chapter introduces BCIs, the annual awards and books, and the twelve projects nominated for a BCI Research Award in 2020.

2 What Is a BCI?

The most widely cited review of BCIs states: "A BCI is a communication system in which messages or commands that an individual sends to the external world do not pass through the brain's normal output pathways of peripheral nerves and muscles. For example, in an EEG based BCI the messages are encoded in EEG activity. A BCI provides its user with an alternative method for acting on the world [7]." While different authors use different terms, most people agree with this general definition.

Most BCIs help people with severe movement disability by replacing or restoring lost movements [8]. The "replace" function includes BCIs for control of prosthetic limbs or spelling systems [1, 2]. Thus, people who can no longer perform abilities like grasping, typing, or speaking can replace lost functions by directly controlling a device with brain activity. Other BCIs help restore lost movements, such as helping people regain upper-limb control after a stroke [3, 4, 6]. In addition to improving voluntary movement control, many patients who participate in stroke rehabilitation therapy using BCIs report reductions in spasms and pain.

The "restore" and "replace" applications require using the BCI at different times. A BCI system to replace lost functions should be available on-demand, even in the middle of the night, without a medical expert or technician. BCIs to restore functions are instead typically used through therapy sessions with a licensed therapist. A common schedule would be 25 sessions with 2–3 sessions per week, typically during regular business hours. While both "replace" and "restore" BCIs must provide real-time feedback to the user, as required by all BCIs, the immediate feedback is the main goal of a "replace" BCI. In a "restore" BCI, the goal is instead to produce long-lasting change.

BCIs are also becoming more prominent as consumer products rather than tools for patients. Facebook, Elon Musk, and others have announced high-profile BCI projects that primarily aim toward consumer devices. Many new companies have emerged that now sell BCIs meant for consumers. These new directions could accelerate BCI research and development, but need to be presented to the public accurately and fairly.

The BCI Research Awards have been informing readers about BCIs for a decade. We hope the awards and books have also encouraged high-quality BCI projects and publications.

3   The Annual BCI Research Award

The BCI Award Foundation organizes the Annual BCI Research Award. Drs. Christoph Guger and Dean Krusienski are both presidents of the BCI Award Foundation, which was founded in 2017. Figure 1 shows the two presidents and the five other Board Members of the BCI Award Foundation.

The BCI Research Award is open to researchers or teams (excluding members of the jury) around the world. Different projects have involved various combinations of hardware, software, algorithms, methods, and other components. The prizes were provided by the Austrian company g.tec medical engineering (author CG is the CEO), IEEE Brain, the BCI Society and by the German company CorTec.

The awards procedure this year followed a procedure like prior years:

1.   The BCI Award Foundation selects a Chairperson of the Jury from a top BCI research institute.

Christoph Guger (AT)
President
g.tec medical engineering GmbH

Dean J. Krusienski (US)
Co-President
Virginia Commonwealth University

Tomasz M. Rutkowski (JP)
Treasurer
RIKEN AIP

Mikhail Lebedev (US)
Ambassador
Duke University Medical Center

Jing Jin (CN)
Ambassador
East China University of Science and Technology

Nuri Firat Ince (US)
Ambassador
University of Houston

Brendan Allison (US)
Chief Editor, Book Series
University of California San Diego

Fig. 1 The BCI Award Foundation has seven board members. All of the board members have been active in BCI research for at least ten years

Aysegul Gunduz (US)
Chair 2020
University of Florida

Sergey D. Stavisky (US)
Winner 2019
Stanford University

Adriane Randolph (US)
Kennesaw State University

Steve Meng (CN)
Shanghai Jiao Tong University

Jörn Rickert (DE)
CorTec GmbH

Fabien Lotte (FR)
INRIA University

Yannick Roy (CA)
NeuroTechX

Fig. 2 The jury for the 2020 BCI Research Award

2. The Chairperson selects a jury of international BCI experts to evaluate all projects submitted for the Award.
3. The Award website[1] announces instructions, scoring criteria, and the submission deadline for projects.
4. After the deadline, the jury members judge each submitted project.
5. The jury chooses the nominees and the first, second, and third place winners.
6. The Award website announces the nominees, and we invite them to that year's Awards Ceremony within a major BCI conference.
7. At the Awards Ceremony, we announce the winners, provide prizes, and thank the jury and the conference organizers.

Initially, the projects that the jury received by the submission deadline were two-page project descriptions. In 2018, we began requiring a supplemental two-minute video about the project. Thus, the jury has had more information to consider when deciding the award, but more work as well (Fig. 2).

The head of the jury in 2020 was Professor Aysegul Gunduz from University of Florida. The 2020 jury included Sergey D. Stavisky, who won the 2019 BCI Research Award. Aysegul Gunduz says: "The 2020 jury also had a good mix of people with backgrounds in invasive and non-invasive BCIs who work in different areas active in BCIs. This prior experience and breadth are both important in juries, who need to evaluate a wide range of BCI projects each year". The 2020 jury was:

---

[1] https://www.bci-award.com/Home.

The scoring criteria that the jury used to select the nominees and winners were the same as all previous BCI Research Awards:

- Does the project include a novel application of the BCI?
- Is there any new methodological approach used compared to earlier projects?
- Is there any new benefit for potential users of a BCI?
- Is there any improvement in terms of speed of the system (e.g. bit/min)?
- Is there any improvement in terms of accuracy of the system?
- Does the project include any results obtained from real patients or other potential users?
- Is the used approach working online/in real-time?
- Is there any improvement in terms of usability?
- Does the project include any novel hardware or software developments?

After all of the jury members have scored all of the projects, the twelve projects with the highest scores are nominated for an award. These nominations are announced through the BCI Award website and other means, and we invite the nominees to the Awards Ceremony. Last year, the 2019 Awards Ceremony was part of the bi-annual BCI Conference in Graz, which also hosted the awards in 2011, 2014, and 2017. For the 2020 Awards, we couldn't schedule an in-person Awards Ceremony due to COVID. So, we scheduled the first online Awards Ceremony. Nominees were invited to participate virtually, including an oral presentation for the Awards Ceremony.

The 2020 Award Ceremony occurred online as part of the virtual IEEE Systems, Man, and Cybernetics conference[2] organized through Toronto from October 11–14, 2020. This conference had a lot of other BCI-related activity, including several BMI sessions and a BR4IN.IO hackathon. The Award Ceremony occurred during the lunch break on October 13, keeping with the tradition of hosting our award ceremonies in a casual atmosphere with food and drinks. The Ceremony included a friendly introduction by Dr. Guger, announcement of the nominees and winner, and videos or statements from some nominees (including brief interviews).

So, although we are still new to hosting the BCI Awards Ceremonies online, we kept other traditions that we've established for these ceremonies. The 2020 Awards Ceremony was successful, and the 2021 Awards Ceremony will be online as well.

The awards for winning projects were also consistent with the ceremonies from the last few years. The jury was asked to choose winners for first, second, and third place. We announced that these winners would receive $3000, $2000, and $1000, respectively, in addition to a certificate and other prizes. The authors of the winning project were also asked to contribute to this book by writing a chapter, and/or joining us for an interview, about their project and related work.

---

[2] http://smc2020.org/.

4   The BCI Research Award Book Series

The preceding section included the timeline for the annual award that ended with the announcement of the winners at the Awards Ceremony. That's when the work on these books begins:

1. We tell the nominees what we need in each chapter, which may be a project summary and/or interview. We encourage images, reports about their newest research future directions, and discussion in addition to project details.
2. Interview some nominees and develop each interview into a chapter.
3. Read the project summary chapters when the authors submit them to us.
4. Edit both types of chapters as needed. Aside from fixing mistakes, we want chapters to be informative and clear.
5. Correspond with the authors to ask for clarification or new text, check on changes, get new references or figures, check copyright issues, and other details.
6. Develop the introduction and conclusion chapters.
7. Submit all chapters to the publisher. (This is where we are as of this writing in April 2021).
8. The publisher then reviews the chapters and sends them to a typesetter.
9. A few months later, the typesetter sends us the proofs, which we share with the chapter authors.
10. Submit any corrections to the proofs and ask the publisher to finalize the book.

The book is always edited by three people: Drs. Guger and Allison from the BCI Award Foundation, and the chair of the jury from that year. This year, our jury chair and co-author is Prof. Aysegul Gunduz from the University of Florida. These three people are also responsible for the introduction and discussion chapters. In practice, Dr. Allison writes most of these two chapters and is primarily responsible for editing the other chapters. Dr. Guger has so far conducted all of the interviews, which Dr. Allison edited into chapters.

We review the chapters and do have the option of rejecting any chapters. However, chapter rejections are rare in the BCI Research Award book series, like other edited books of this nature. The underlying projects already succeeded in a very challenging peer-review process through the jury selection procedure. Thus, the authors are capable of producing high-quality material describing a good BCI project that they usually want to publicize.

This is only the second book that includes interview chapters. This year, we have seven project summaries and four interviews from the nominees, who are in the next section. In our book for the 2018 Awards, we included the interviews within the discussion chapter [5]. In 2019, we tried expanding the interviews with more questions and supporting components such as an introduction, images, and references. These interviews led to good chapters that were easier to read than project summary chapters. We aim to make interview chapters easier to read than project summary chapters to provide different levels of difficulty for different readers.

5 Projects Nominated for the BCI Award 2020

The twelve submissions with the highest scores were nominated for the BCI Research Award 2020. These twelve projects, followed by authors and affiliations, were:

Enhancing Gesture Decoding Performance Using Signals from Human Posterior Parietal Cortex

Guangye Li[1], Meng Wang[1], Shize Jiang[2], Jie Hu[2], Liang Chen[2], Dingguo Zhang[3]

1 *Robotics Institute, School of Mechanical Engineering, Shanghai Jiao Tong University, Shanghai, China*
2 *Neurosurgery Department of Huashan Hospital, Fudan University, Shanghai, China*
3 *Department of Electronic and Electrical Engineering, University of Bath, Bath, UK*

Machine Translation of Cortical Activity to Text

Joseph G. Makin, David A. Moses, Edward F. Chang

Center for Integrative Neuroscience/Department of Neurological Surgery, University of California, San Francisco

Towards Practical MEG-BCI with Optically Pumped Magnetometers

Benjamin Wittevrongel[1,3], Niall Holmes[2], Elena Boto[2], Ryan Hill[2], Molly Rea[2], Ben Hunt[2], Arno Libert[1], Elvira Khachatryan[1], Marc Van Hulle[1,3], Richard Bowtell[2], Matthew Brookes[2]

1 *Laboratory for Neuro- and Psychophysiology, KU Leuven, Belgium*
2 *Department of Physics and Astronomy, University of Nottingham, UK*
3 *KU Leuven institute for Artificial Intelligence (Leuven.AI), Belgium*

EEG Decoding of Pain Perception for a Real-Time Reflex System in Prostheses

Zied Tayeb[1], Rohit Bose[2,3], Andrei Dragomir[2,4], Luke E. Osborn[5,6], Nitish V. Thakor[5,7], Gordon Cheng[1]

1 *Institute for Cognitive Systems, Technical University of Munich, Germany*
2 *Institute for Health, National University of Singapore, Singapore*
3 *Department of Bioengineering, University of Pittsburgh, USA*
4 *Department of Biomedical Engineering, University of Houston, USA*
5 *Department of Biomedical Engineering, Johns Hopkins School of Medicine, USA*

6 Research Exploratory Development, Johns Hopkins University Applied Physics Laboratory, USA

7 Department of Biomedical Engineering, National University of Singapore, Singapore

A Computer-Brain Interface that Restores Lost Extremities Touch and Movement Sensations

Giacomo Valle, Francesco Maria Petrini Pavle Mijovic, Bogdan Mijovic, Stanisa Raspopovic

Institute for Robotics and Intelligent Systems, ETH Zürich

Restoring the Sense of Touch Using a Sensorimotor Demultiplexing Neural Interface

Patrick D. Ganzer, Samuel C. Colachis, Michael A. Schwemmer, David A. Friedenberg, Collin F. Dunlap, Carly E. Swiftney, Adam F. Jacobowitz, Doug J. Weber, Marcia A. Bockbrader, Gaurav Sharma

Battelle Memorial Institute, USA

A Brain–Spine Interface Complements Deep-Brain Stimulation to Both Alleviate Gait and Balance Deficits and Increase Alertness in a Primate Model of Parkinson's Disease

Tomislav Milekovic[1,2,3,4], Flavio Raschellà[1,2,3,5], Matthew G. Perich[2], Eduardo Martin Moraud[1,2,3,6], Shiqi Sun[1,2,3,7], Giuseppe Schiavone[8], Yang Jianzhong[9,10], Andrea Galvez[1,2,3,4], Christopher Hitz[1], Alessio Salomon[1], Jimmy Ravier[1,2,3], David Borton[1,11], Jean Laurens[1,12], Isabelle Vollenweider[1], Simon Borgognon[1,2,3], Jean-Baptiste Mignardot[1], Wai Kin D Ko[9,10], Cheng YunLong[9,10], Li Hao[9,10], Peng Hao[9,10], Laurent Petit[13,14], Qin Li[9,10], Marco Capogrosso[1], Tim Denison[15], Stéphanie P. Lacour[8], Silvestro Micera[5,16], Chuan Qin[10], Jocelyne Bloch[1,2,3,6], Erwan Bezard[9,10,13,14], Grégoire Courtine[1,2,3,6]

1 Center for Neuroprosthetics (CNP) and Brain Mind Institute, School of Life Sciences, Swiss Federal Institute of Technology (EPFL), Switzerland
2 Department of Clinical Neuroscience, Lausanne University Hospital (CHUV) and University of Lausanne (UNIL), Switzerland
3 Defitech Center for Interventional Neurotherapies (NeuroRestore), CHUV/UNIL/EPFL, Switzerland
4 Department of Fundamental Neuroscience, Faculty of Medicine, University of Geneva, Switzerland
5 CNP and Institute of Bioengineering, School of Engineering, EPFL, Switzerland
6 Department of Neurosurgery, CHUV, Switzerland

7 *Beijing Engineering Research Center for Intelligent Rehabilitation, College of Engineering, Peking University, People's Republic of China*
8 *CNP, Institute of Microengineering and Institute of Bioengineering, School of Engineering, EPFL, Switzerland*
9 *Motac Neuroscience, UK*
10 *Institute of Laboratory Animal Sciences, China Academy of Medical Sciences, People's Republic of China*
11 *Carney Institute for Brain Science, School of Engineering, Brown University, USA*
12 *Department of Neuroscience, Baylor College of Medicine, USA*
13 *Université de Bordeaux, Institut des Maladies Neurodégénératives (IMN), UMR 5293, France*
14 *CNRS, IMN, UMR 5293, France*
15 *Oxford University, UK*
16 *The BioRobotics Institute, Scuola Superiore Sant'Anna, Italy*

Speaker-Independent Auditory Attention Decoding Without Access to Clean Speech Sources

Cong Han[1,2], James O'Sullivan[1,2], Yi Luo[1,2], Jose Herrero[3], Ashesh D. Mehta[3], Nima Mesgarani[1,2]

1 *Department of Electrical Engineering, Columbia University, New York, NY, USA*
2 *Zuckerman Mind Brain Behavior Institute, Columbia University, New York, NY, USA*
3 *Department of Neurosurgery, Hofstra-Northwell School of Medicine and Feinstein Institute for Medical Research, Manhasset, New York, NY, USA*

A High-Performance Handwriting BCI

Francis R. Willett[1,2], Donald T. Avansino[1], Leigh Hochberg[3], Jaimie Henderson[1], Krishna V. Shenoy[1,2]

1 *Stanford University, USA*
2 *Howard Hughes Medical Institute, USA*
3 *Brown University, Harvard Medical School, Massachusetts General Hospital, USA*

A Neuromorphic Brain Computer Interface for Real-Time Detection of a New Biomarker for Epilepsy Surgery

Karla Burelo[1,2], Mohammadali Sharifhazileh[1,2], Johannes Sarnthein[2], and Giacomo Indiveri[1]

1 *University of Zurich and ETH Zurich, Institute of Neuroinformatics, Switzerland*
2 *University Hospital and University of Zurich, Switzerland*

"Sono-Optogenetics": An Ultrasound-Mediated Non-invasive Optogenetic Brain-Computer Interface

Xiang Wu, Paul Chong, Huiliang Wang, Guosong Hong

Department of Materials Science and Engineering, Wu Tsai Neurosciences Institute, Stanford University, USA

High-Dimensional (8D) Control of Complex Effectors Such as an Exoskeleton by a Tetraplegic Subject Using Chronic ECoG Recordings Using Stable and Robust Over Time Adaptive Direct Neural Decoder

Alexandre Moly[1], Thomas Costecalde[1], Félix Martel[1], Antoine Lassauce[1], Serpil Karakas[1], Gael Reganha[1], Alexandre Verney[2], Benoit Milville[2], Guillaume Charvet[1], Stéphan Chabardes[3], Alim Louis Benabid[1], Tetiana Aksenova[1]

1 *CEA, LETI, CLINATEC, University Grenoble Alpes, MINATEC, France*
2 *CEA, LIST, DIASI, SRI, Gif-sur-Yvette, France*
3 *Centre Hospitalier Universitaire Grenoble Alpes, France*

This year, for the first time, we posted videos from half of the nominated projects on the BCI Award website.[3] Each video was developed by the team behind the project submission. Aside from learning more about the projects, you can also see and hear some of the people behind each project and get a sense of what different BCI research labs look like. Most videos last about two minutes and include:

- Clips of the BCI system in operation;
- Graphics, animations, and text to illustrate system components, procedures, and project results;
- Commentary from patients and project developers;
- Logos from the institutes where projects were executed; and/or
- Supporting references.

6 Summary

This book is about the Tenth Annual BCI Research Awards. The next several chapters feature project descriptions and interviews based on the projects that were nominated for a BCI Research Award this year. Each chapter addresses how the system measures information from the brain, including different types of implanted and non-implanted approaches, such as EEG, ECoG, or MEG. Each project also describes the signal

---

[3] https://www.bci-award.com/2020.

translation and outputs, such as a speller on a monitor, prosthetic limb, or exoskeleton. Most chapters report BCIs for patient applications, including BCI systems that could restore touch, movement, communication, or freedom from epilepsy.

The concluding chapter presents the winners of the 2020 BCI Research Awards, shares some information about next year's awards, and features some discussion. Next year's awards will feature some changes in sponsors and other minor details, along with the new jury. However, we will not change the submission procedure, award criteria, or number of nominees and winners in the Eleventh BCI Research Award.

References

1. Abiri R, Borhani S, Sellers EW, Jiang Y, Zhao X (2019) A comprehensive review of EEG-based brain–computer interface paradigms. J Neural Eng 16(1):011001
2. Allison BZ, Kübler A, Jin J (2020) 30+ years of P300 brain-computer interfaces. Psychophysiology e13569–e13569
3. Biasiucci A, Leeb R, Iturrate I, Perdikis S, Al-Khodairy A, Corbet T, Millán JDR (2018) Brain-actuated functional electrical stimulation elicits lasting arm motor recovery after stroke. Nat Commun 9(1):1–13
4. Cervera MA, Soekadar SR, Ushiba J, Millán JDR, Liu M, Birbaumer N, Garipelli G (2018) Brain-computer interfaces for post-stroke motor rehabilitation: a meta-analysis. Ann Clin Transl Neurol 5(5):651–663
5. Guger C, Allison BZ, Miller K (2020) Highlights and interviews with winners. In: Brain–computer interface research. Springer, Cham, pp 107–121
6. Sebastián-Romagosa M, Cho W, Ortner R, Murovec N, Von Oertzen T, Kamada K et al (2020) Brain computer interface treatment for motor rehabilitation of upper extremity of stroke patients—a feasibility study. Front Neurosci 14
7. Wolpaw JR, Birbaumer N, McFarland DJ, Pfurtscheller G, Vaughan TM (2002) Brain–computer interfaces for communication and control. Clin Neurophysiol 113(6):767–791
8. Wolpaw JR, Wolpaw EW (2012) Brain-computer interfaces: something new under the sun. Brain-Comput Interfaces Principles Pract 14

Evaluation of Human Posterior Parietal Cortex in Gesture Decoding Performance Enhancement Using Stereoelectroencephalography (SEEG) Signals

Guangye Li, Meng Wang, Shize Jiang, Jie Hu, Liang Chen, and Dingguo Zhang

Abstract Previous intracranial electroencephalography (iEEG)-based brain-machine interfaces (BMIs) towards gesture decoding mostly used neural signals from the primary sensorimotor cortex while largely ignoring the hand movement related signals from posterior parietal cortex (PPC). In this work, we investigated the role of human PPC during a three-class hand gesture task using stereoelectroen-cephalography (SEEG) recordings from 25 subjects. Using the high gamma power (55–150 Hz) of SEEG signal recorded within three ROIs [PPC, postcentral cortex (POC) and precentral cortex (PRC)], we computed four indices for each of ROI, including: (1) activation strength; (2) gesture selectivity; (3) first activation time; (4) decoding accuracy. We find that a majority (L: 60%, R: 40%) of electrodes in all three ROIs present significant activation during the task. The activation of PPC, from a large temporal scale, is earlier than the sensorimotor cortex (PRC and POC). Among the activated electrodes, 15% (PRC), 26% (POC) and 4% (left PPC) of electrodes are significantly selective to gestures. Finally, decoding accuracy obtained by combining the selective electrodes from PPC with the sensorimotor cortex together is 5% higher than that from sensorimotor cortex only. Above all, our results suggest that PPC could be a rich neural source for iEEG-based BMI. The early activation of PPC may provide additional implications for further scientific research and high-level BMI applications.

Keywords Brain-machine interface · Gesture decoding · Posterior parietal cortex · SEEG

G. Li · M. Wang
School of Mechanical Engineering, Robotics Institute, Shanghai Jiao Tong University, Shanghai, China

S. Jiang · J. Hu · L. Chen
Neurosurgery Department, Huashan Hospital, Fudan University, Shanghai, China

D. Zhang (✉)
Department of Electronic and Electrical Engineering, University of Bath, Bath, UK
e-mail: d.zhang@bath.ac.uk

© The Author(s), under exclusive license to Springer Nature Switzerland AG 2021
C. Guger et al. (eds.), *Brain-Computer Interface Research*,
SpringerBriefs in Electrical and Computer Engineering,
https://doi.org/10.1007/978-3-030-79287-9_2

1 Introduction

Human intracranial electroencephalography (iEEG) recordings [e.g., electrocorticography (ECoG) and stereoelectroencephalography (SEEG)] hold the advantage of stably recording rich neural information across multiple brain regions, and iEEG-based brain-machine interfaces (BMIs) have made significant achievements in decoding frequently-used functional gestures [1–4]. Nevertheless, most of these iEEG studies mainly focus neural signals from primary sensorimotor cortex (i.e., precentral and postcentral cortex, abbr. as PRC and POC respectively) to decode gestures, which works well but may not be optimal.

Beyond primary sensorimotor cortex, posterior parietal cortex (PPC) has been reported to be related with multiple hand movements like grasp, intransitive postures and even pantomime and imagination of those movements; moreover, they may contribute to the formation of early motor intention [5, 6]. PPC lesions cause severe deficits of controlling hand shape appropriate for objects [7]. Intracortical electrical stimulation to the lateral region of monkey's area 5 within PPC elicits finger and wrist movements [8]. Functional magnetic resonance imaging (fMRI) studies reveal that grip type (power vs. precision grasp) is encoded in several subareas of PPC [9, 10]. A human electrophysiological study on a tetraplegic patient by Klaes et al. has also shown that the imagination of scissor-rock-paper hand gestures can be decoded by spiking activity in PPC [11]. Taken together, all these works support the notion that the PPC has distinct relevance to hand movement control.

However, although electrophysiological studies have demonstrated that spiking activity from PPC is involved in hand shape encoding, whether iEEG recordings from human PPC could also benefit gesture decoding is still unclear. In this work, we conducted research using human SEEG recordings to address this question.

2 Methods

Signals, Electrodes, Subjects and Experiments: SEEG signals from 25 epilepsy subjects were recorded in hospital using a clinical recording system (EEG-1200C, Nihon Kohden, Irvine, CA) sampled at 500–2000 Hz. The 25 subjects had a total of 3501 contacts implanted (each contact: 2.0 mm long, 0.8 mm diameter, center-center distance 3.5 mm). We then localized the position of each contact using post-implantation CT images with pre-implantation MRI images with the help of iEEGview toolbox [12]. The anatomical label of each electrode contact was identified by cortical parcellation and subcortical segmentation results under the Desikan-Killiany atlas [13, 14]. Finally, each electrode was mapped from the individual brain to a standard MNI (Montreal Neurological Institute) brain template for the purpose of group analysis. Since we were interested in the neural response in PRC, POC and PPC, only electrodes within these regions of interest (ROIs) were included in this study. As a result, 221, 114 and 524 SEEG contacts were identified in PRC, POC and PPC across all subjects (Fig. 1b). The subjects repeatedly performed the hand movement task using the hand contralateral to the hemisphere with the majority of the implanted SEEG electrodes. Figure 1a shows the experimental protocol of this study and the experiment details. During the recording, two surface EMG signals taken from the extensor carpi radialis muscle were recorded simultaneously using the same equipment with SEEG. The study was approved by the Ethics Committee of Huashan Hospital (Shanghai, China).

Data Processing: For each subject, the raw signals were first resampled to 1000 Hz using the built-in Matlab function (*resample*) to facilitate consistent computation across subjects. Second, the channels with line noise power at 50 Hz larger than a significance level were removed from further analysis. The significance level was defined as the summation of median line noise power across all electrodes and 10 times of their median absolute deviation. Third, the 50 Hz line noise and its harmonics were removed using a comb notch filter. Fourth, the signals were re-referenced using a Laplacian scheme to further improve the signal quality [15]. Finally, we extracted the high gamma power at 55–150 Hz using the Hilbert transform. The derived high gamma power was then divided into 100-ms time bins for subsequent analysis. Additionally, we also identified the movement onset by finding the time point when the absolute amplitude of EMG first time exceeded an adaptive threshold using the

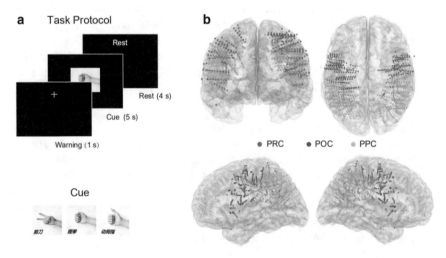

Fig. 1 Task protocol and the distribution of SEEG contacts from three ROIs. **a** The protocol started from a 1-s warning phase to alert the subject in each trial. Next, a gesture picture appeared for 5 s (Cue phase) to instruct the subject to repeatedly perform the corresponding gesture. Following that was a 4-s rest phase during which the subject relaxed without any movement. The task used three hand gestures (scissor, rock and thumb). One of three gestures was randomly selected and displayed in each trial, and each subject performed 20 trials for each gesture. **b** Front/top/left/right views of the locations of SEEG contacts in a standard MNI (Montreal Neurological Institute) brain template. All contacts within ROIs from 25 subjects are shown. Each colored dot represents one SEEG contact in a specific ROI

methods described in [16]. To be concise, the movement onset was denoted as *Move* and the time when the Cue (Fig. 1a) appeared was denoted as *Cue* throughout this article.

Performance Index: Using high gamma power activity, we evaluated the characteristics of neural responses and possible functions from these ROIs (PRC, POC and PPC) in relation to the hand movement through four performance indices:

1. ***Activation Strength***, which measures the degree of activation during task compared to rest and reflects the general involvement in the task. The activation strength was computed with the coefficient of determination r^2 in Eq. (1) using the movement state ($c = 2$, i.e., task/rest).

$$r^2 = \frac{\sum_{i=1}^{c} m\left(\overline{X_i} - \overline{X}\right)^2}{\sum_{i=1}^{c} \sum_{j=1}^{m} \left(X_{i,j} - \overline{X}\right)^2} \tag{1}$$

where $\overline{X_i} = \sum_{j=1}^{m} X_{i,j}$ and $\overline{X} = \sum_{i=1}^{c} \sum_{j=1}^{m} X_{i,j}$. $X_{i,j}$ indicates the high gamma power from *ith* class and *jth* trial; c: number of classes, m: number of trials. After obtaining the observed r^2, we then implemented a permutation test [15] with 3000 repeats to determine the significance of activation where the class (i.e., rest and task) was randomly permuted across trials and r^2 was recomputed, generating a distribution of surrogate r^2 and revealing p values of the observed r^2. The above calculation was performed at each electrode and window. Any electrode was called an activated electrode if its q value obtained after FDR correction was smaller than 0.001 at any time window within $[-0.5, 0.5]$ s around *Move*.

2. ***Gesture Selectivity***, which gauges the difference of the response to three gestures. The computation process was similar to (1), but instead of using the movement state ($c = 2$, i.e., task/rest) as classes, here three different gestures ($c = 3$, i.e., scissor, rock and thumb, Fig. 1a) were used as classes in Eq. (1). Similar to (1), a random permutation test (3000 times) was conducted to identify the significantly selective electrodes for each subject. Any electrode was called a selective electrode if its q value obtained after FDR correction was smaller than 0.05 at any time window within $[-0.5, 0.5]$ s around *Move*.

3. ***First Activation Time***, indicating the first time point when an electrode presented significant activation strength in the task. In detail, the first activation time of an electrode was defined as the first time when r^2 (activation strength) during $[-0.5, 0.5]$ s around *Move* presented significance ($p < 0.001$).

4. ***Decoding Accuracy***, which evaluates the gesture decoding performance. The electrodes that present gesture selectivity within each ROI were used for the calculation of decoding accuracy. Due to the limited samples in current study, forward optimal feature selection (fOFS) was applied first to reduce the feature

dimension and thus avoid overfitting. Specifically, starting from emptiness, the optimal feature vector successively added the best feature that could maximize the decoding accuracy (DA) when included into the current optimal feature vector, until DA stopped increasing with further additions. The DA was the average over five accuracy values that were obtained by linear support vector machine (SVM) with fivefold cross validation. To avoid bias of the finally selected optimal feature set, we repeated the above-mentioned whole fOFS procedure 100 times with 100 different partitions, which yielded 100 nonidentical optimal feature sets and thus 100 DAs. The average of these DAs was used as the final DA for each subject.

3 Results

Among all the SEEG contacts from these three ROIs, 67% (n = 147, L = 93, R = 54), 60% (n = 68, L = 49, R = 19) and 51% (n = 265, L = 200, R = 65) got activated during tasks in PRC, POC and PPC respectively, indicating a majority of electrodes within all three ROIs were involved in the task (Fig. 2a). Moreover, among the activated contacts, 15% (PRC, bilateral average), 26% (POC, bilateral average) and 4% (PPC, left) presented significant selectivity for gestures. This activation result indicated that a relatively small group of electrodes were selective among the activated electrodes for PRC and POC, while a minority were selective for PPC (Fig. 2b). The selective electrodes within each ROI at either hemisphere came from 12% (n = 3 (rounded average), L = 4, R = 1) subjects on average.

Within these ROIs, the median first activation time was about 310 ms for PPC, 400 ms for PRC and 510 ms for POC respectively the stimulus onset. The activation of ROIs was significantly (one side Wilcoxon Ranksum test with Bonferroni correction)

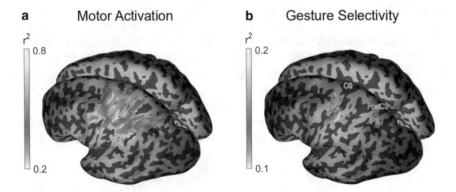

Fig. 2 Distribution of task-related information from three ROIs. **a, b** The spatial distribution of activation information from recorded PRC, POC and PPC electrodes. The maximum r^2 value over time is used to generate the map (See the section titled "Performance Index"). For visualization purposes, results were rendered on the left hemisphere only

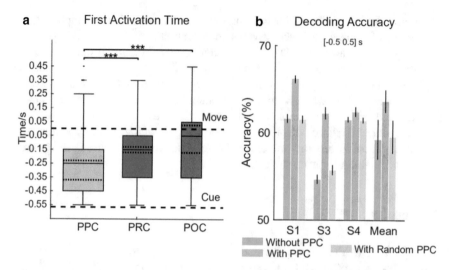

Fig. 3 First activation time results and the effect of PPC in decoding performance. **a** First activation time. Inside each box, the black solid line denotes the median first activation time (q_2). The lower and upper boundary of the box denotes the 25% (q_1) and 75% (q_3) percentile. The width of each box is scaled to the number of samples. The whisker of each box extends to $q_3 + 1.5 * (q_3 - q_1)$. *Move* indicates the average EMG onset (0 s). ***, $p < 0.001$ (one side Wilcoxon rank sum test after Bonferroni correction), $p = 0.06$ for PRC and POC. **b** Decoding performance. *Blue*: The average decoding accuracy using gesture selective electrodes located at PRC and POC only. *Orange*: Similar to *Blue,* but PPC is also included. *Gray*: Similar to *Orange,* but the temporal sequence of the features from PCC is randomized across trials. The error bar indicates the standard error. The computation shows results from three subjects. Mean values are computed across these three subjects. The time shown at the top of the figure indicates the time period used for the calculation. Time 0 is the same with **a**

sequential along time course, where PPC activated first, PRC second and POC last (Fig. 3a). Importantly, combining spatial features from PPC improved the decoding accuracy (DA) by 5% on average compared to using only primary sensorimotor cortex (Fig. 3b). As a comparison, when randomizing the PPC features across trials and repeating the decoding process (Fig. 3b, With Random PPC), no improvement on the DA was seen, indicating that the PPC contained useful visuomotor information that can assist the gesture decoding.

In summary, our results suggest that human PPC encodes specific information about fine hand movements that is complementary to that of primary sensorimotor cortex, potentially providing a new signal source that will benefit further iEEG-based BCI applications.

4 Discussion

In current study, we evaluated whether human PPC encoded neural information related with hand gestures; moreover, we investigated whether PPC together with primary sensorimotor cortex (PRC and POC) could enhance the hand gesture decoding using SEEG recordings from 25 human subjects participating in a hand gesture experiment. We proposed four performance indices and aimed to answer a series of relevant critical questions: (1) to what extent PPC is involved to the task; (2) whether a temporal activation sequence exists between PPC and primary sensorimotor cortex during the visuomotor task; (3) whether iEEG recordings from PPC contain information about fine gestures and can subsequently improve decoding performance.

Incorporating LFP recordings across multiple subjects, we have found that a majority of electrodes located in PPC, PRC and POC are activated with a temporal sequence in terms of HGP, where PPC activates first, PRC second and POC last (Figs. 2a and 3a). Two other human iEEG studies present similar results [17, 18]. These two studies did not directly address the same question as this work, but also used HGP and conducted similar behavior tasks. Figure 2 of [17] shows that three ECoG electrodes located at PPC, PRC and POC are sequentially activated around 250 ms, 400 ms and 590 ms after *Cue* with an EMG onset at 576 ms, similar to 310 ms, 400 ms, 510 ms after *Cue* with EMG onset at 564 ms in current study. The results in this work may indicate that, in a large scale, each ROI has a relative temporal sequence during the task information processing.

The LFP results in this work show that human PPC is selective to hand gestures (Fig. 2b), verifying our hypothesis that human PPC iEEG recordings contain gesture related information, which is consistent with previous monkey LFP studies and human spike studies [11, 19]. The small group of PPC selective electrodes can still provide effective information for classification and such information is complementary to the primary sensorimotor area (Fig. 3b).

Notably, this work did not answer the question of which role the PPC is played in the visuomotor task. Early reports show that PPC contains visual-dominant, motor-dominant and visuomotor neurons [20–22]. However, whether PPC is more related to the visual shape of objects or the hand shape is still in debate. Therefore, further studies should be conducted to investigate this question.

Acknowledgements This work was supported by grants from the National Natural Science Foundation of China (No. 91848112, No. 61761166006), the Shanghai Municipal Commission of Health and Family Planning (No. 2017ZZ01006), and the Shanghai Municipal Science and Technology Major Project (No. 2018SHZDZX03).

References

1. Parvizi J, Kastner S (2018) Promises and limitations of human intracranial electroencephalography. Nat Neurosci 21:474–483
2. Yanagisawa T, Hirata M, Saitoh Y, Kishima H, Matsushita K, Goto T, Fukuma R, Yokoi H, Kamitani Y, Yoshimine T (2012) Electrocorticographic control of a prosthetic arm in paralyzed patients. Ann Neurol 71:353–361
3. Branco MP, Freudenburg ZV, Aarnoutse EJ, Bleichner MG, Vansteensel MJ, Ramsey NF (2017) Decoding hand gestures from primary somatosensory cortex using high-density ECoG. Neuroimage 147:130–142
4. Li G, Jiang S, Xu Y, Wu Z, Chen L, Zhang D (2017) A preliminary study towards prosthetic hand control using human stereo-electroencephalography (SEEG) signals. In: 2017 8th international IEEE/EMBS conference on neural engineering (NER), pp 375–378
5. Andersen RA, Buneo CA (2002) Intentional maps in posterior parietal cortex. Annu Rev Neurosci 25:189–220
6. Vingerhoets G (2014) Contribution of the posterior parietal cortex in reaching, grasping, and using objects and tools. Front Psychol 5
7. Jeannerod M, Decety J, Michel F (1994) Impairment of grasping movements following a bilateral posterior parietal lesion. Neuropsychologia 32:369–380
8. Rathelot JA, Dum RP, Strick PL (2017) Posterior parietal cortex contains a command apparatus for hand movements. Proc Natl Acad Sci USA 114:4255–4260
9. Di Bono MG, Begliomini C, Castiello U, Zorzi M (2015) Probing the reaching-grasping network in humans through multivoxel pattern decoding. Brain Behav 5:e00412
10. Fabbri S, Stubbs KM, Cusack R, Culham JC (2016) Disentangling representations of object and grasp properties in the human brain. J Neurosci 36:7648–7662
11. Klaes C, Kellis S, Aflalo T, Lee B, Pejsa K, Shanfield K, Hayes-Jackson S, Aisen M, Heck C, Liu C, Andersen RA (2015) Hand shape representations in the human posterior parietal cortex. J Neurosci official J Soc Neurosci 35:15466–15476
12. Li G, Jiang S, Chen C, Brunner P, Wu Z, Schalk G, Chen L, Zhang D (2019) iEEGview: an open-source multifunction GUI-based Matlab toolbox for localization and visualization of human intracranial electrodes. J Neural Eng 17:016016
13. Fischl B, Salat DH, Busa E, Albert M, Dieterich M, Haselgrove C, van der Kouwe A, Killiany R, Kennedy D, Klaveness S, Montillo A, Makris N, Rosen B, Dale AM (2002) Whole brain segmentation. Neuron 33:341–355
14. Desikan RS, Ségonne F, Fischl B, Quinn BT, Dickerson BC, Blacker D, Buckner RL, Dale AM, Maguire RP, Hyman BT, Albert MS, Killiany RJ (2006) An automated labeling system for subdividing the human cerebral cortex on MRI scans into gyral based regions of interest. Neuroimage 31:968–980
15. Li G, Jiang S, Paraskevopoulou SE, Wang M, Xu Y, Wu Z, Chen L, Zhang D, Schalk G (2018) Optimal referencing for stereo-electroencephalographic (SEEG) recordings. Neuroimage 327–335
16. Sedghamiz H (2018) BioSigKit: a Matlab toolbox and interface for analysis of biosignals. J Open Source Softw 3:671
17. Coon WG, Gunduz A, Brunner P, Ritaccio AL, Pesaran B, Schalk G (2016) Oscillatory phase modulates the timing of neuronal activations and resulting behavior. Neuroimage 133:294–301
18. Coon WG, Schalk G (2016) A method to establish the spatiotemporal evolution of task-related cortical activity from electrocorticographic signals in single trials. J Neurosci Methods 271:76–85
19. Asher I, Stark E, Abeles M, Prut Y (2007) Comparison of direction and object selectivity of local field potentials and single units in macaque posterior parietal cortex during prehension. J Neurophysiol 97:3684–3695
20. Mountcastle VB, Lynch JC, Georgopoulos A, Sakata H, Acuna C (1975) Posterior parietal association cortex of the monkey: command functions for operations within extrapersonal space. J Neurophysiol 38:871–908

21. Sakata H, Taira M, Murata A, Mine S (1995) Neural mechanisms of visual guidance of hand action in the parietal cortex of the monkey. Cereb Cortex 5:429–438
22. Murata A, Gallese V, Luppino G, Kaseda M, Sakata H (2000) Selectivity for the shape, size, and orientation of objects for grasping in neurons of monkey parietal area AIP. J Neurophysiol 83:2580–2601

Speech Decoding as Machine Translation

Joseph G. Makin, David A. Moses, and Edward F. Chang

Abstract We aimed to improve the state of the art in decoding speech from neural activity, with the ultimate goal of developing a useful brain-machine interface (BMI) for individuals who have lost the ability to speak—from ALS, a stroke, or other traumatic brain injury. In our recent study (Makin et al. in Nat Neurosci 23:575–582, 2020), each of four participants undergoing clinical monitoring for epilepsy read aloud, making repeated passes through a set of some 30–50 sentences, while her electrocorticogram was simultaneously recorded. Our algorithm, which was inspired by recent ideas in machine translation, brought word error rates down from the previous state of the art, about 60, to 3%. In this chapter, we discuss those results, their limitations, and their implications for the general problem of speech decoding.

Keywords Brain-machine interface · ECoG · Speech decoding · Encoder-decoder networks

1 Introduction

The field of speech decoding began in 2009 with the successful synthesis of vowel formants from the firing rates of a small number of neurons, recorded with a micro-electrode implanted into speech-motor cortex of a locked-in patient [3]. Isolated phonemes and monosyllables have subsequently been classified, with moderate accu-

J. G. Makin is now with the School of Electrical and Computer Engineering at Purdue University. For questions about the algorithm/code, contact him at jgmakin@purdue.edu. For questions about experiment/data, contact EFC at edward.chang@ucsf.edu.

J. G. Makin (✉) · D. A. Moses · E. F. Chang
Center for Integrative Neuroscience, UCSF, San Francisco, CA, USA
e-mail: jgmakin@purdue.edu

E. F. Chang
e-mail: edward.chang@ucsf.edu

Department of Neurological Surgery, UCSF, San Francisco, CA, USA

© The Author(s), under exclusive license to Springer Nature Switzerland AG 2021
C. Guger et al. (eds.), *Brain-Computer Interface Research*,
SpringerBriefs in Electrical and Computer Engineering,
https://doi.org/10.1007/978-3-030-79287-9_3

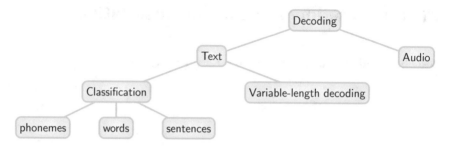

Fig. 1 Strategies for speech decoding

racies, from recordings made by penetrating electrodes [18] or from the electrocorticogram (ECoG) [3, 4, 14, 16]. Even setting aside their modest accuracies, these
results are of interest mostly as proof of concept, because the phonemes produced in
continuous speech are highly influenced by their neighbors ("coarticulation"); and
(taking the other horn of the dilemma) a speech BMI that required its users to produce
phonemes in isolation would forego the principal merits of decoding speech rather
than handwriting or typing: speed and naturalness.

Several studies have attempted to decode continuously spoken speech [1, 2,
9, 11–13]. These can be divided on the basis of what modality they attempt to
decode: audio ("speech synthesis") or text (see Fig. 1). Neural speech synthesis has
markedly improved since the foundational work of Brumberg and colleagues [3],
producing nearly intelligible output. In arguably the most successful of these studies
[2], volunteers were subsequently recruited to transcribe the speech that had been
synthesized from patients' ECoG data, in order to quantify the results. When limited
to a vocabulary of just 50 words, transcribers achieved word error rates[1] (WERs) of
about 50%. (On the other hand, the primary advantage of speech synthesis is that it
is not, in principle, limited to a fixed vocabulary.)

When the aim is, alternatively, to output text, there remains the question of granu-
larity. At one extreme, phonemes can be classified, and subsequently assembled into
words and sentences with a language model. Operating with ECoG and a vocabulary
of 100 words, such an approach has yielded word error rates of about 60% [9]. At
the opposite extreme, Moses and colleagues classified entire sentences from their
corresponding ECoG signatures [13], trading coverage of English for distinguisha-
bility of the tokens to be decoded. This model achieved WERs of 33% on a set of 50
sentences [11].

An alternative to classifying (and subsequently assembling into larger units)
phonemes, words, or sentences is to decode variable-length sequences of words.
That is, the decoder consumes a long sequence of neural data, contemporaneous

---

[1] Errors are computed as the *minimum* number of word insertions, deletions, and substitutions
required to transform the predicted into the true word sequence. Dividing by the number of words
in the true sequence yields a word error *rate*. Intuitively, any sensible decoder should achieve error
rates between 0 and 1.0, since the WER for a "decoder" that just predicts an empty sequence for
every "input" is precisely 1.0. But in practice poor decoders can make errors at rates greater than 1.

with (for example) a single spoken sentence, and then begins emitting words, one at a time, until it decides to stop. The potential advantage of such an approach is that it does not impose assumptions about which parts of the ECoG signal correspond to which words or word parts. This allows for: phoneme classification and assembly into words to be solved jointly rather than sequentially; automatic handling of coarticulation; the assimilation of temporally dispersed (e.g., semantic) information; dispensing with a phoneme transcription, which would in any case be difficult to obtain from non-speaking persons; and the production of any sentence composed from the fixed vocabulary of words. The entire pipeline can be implemented as an artificial neural network, and trained end-to-end, from neural data to sentences. And indeed, such "encoder-decoder" networks have in the last five years become the standard for machine translation, where the input is a variable-length sequence, not of neural data, but of words in another language [7, 8, 19, 21].

Using ECoG as input to an encoder-decoder neural network, we achieved WERs as low as 3%, operating with a vocabulary of about 250 words and a set of 50 unique sentences [11]. Below we reprise those results and discuss their limitations.

2 Methods

We briefly describe the fundamental aspects of the study's methods. More details can be found in the original publication [11].

Participants. Drug-resistant epilepsy can sometimes be treated with brain surgery, in which case seizures are first localized with a neurological recording device. One common procedure is to perform a craniotomy and then place a grid of electrodes on the surface of the brain and monitor the electrocorticogram over the course of (typically) one or two weeks. During this period, patients are not anaesthetized and are able (*inter alia*) to read aloud without difficulty. The participants in the study reviewed here were epilepsy patients at the UCSF Medical Center. Prior to surgery, all participants (four female; all right-handed and left-hemisphere language-dominant; aged 47 [participant **a**], 31 [participant **b**], 29 [participant **c**], and 49 [participant **d**] years) gave written consent to take part in the study, which was carried out according to protocol approved by the UCSF Committee on Human Research.

Data. The electrocorticogram was recorded with high-density (4-mm pitch) arrays from the peri-Sylvian cortices of participants while they read aloud from one of two sets of sentences (see below). The ECoG on all channels and the microphone signal were then pre-processed offline according to the pipelines in Fig. 2a, b, respectively. Finally, the spoken sentences were transcribed. Participants occasionally misread or otherwise misproduced the prompts, so the transcriptions did not always precisely match them. However, the rare (less than one percent of the total) productions that did not correspond to any word in the relevant sentence set (see below) were all transcribed as a single out-of-vocabulary token.

a

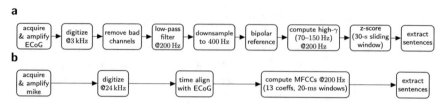

Fig. 2 Data preprocessing

Altogether, these provided a set of data "triplets," each of which consisted of

1. a matrix of ECoG data, with size (number of channels × number of samples);
2. a matrix of audio data (13 MFCCs × number of samples); and
3. the sequence of words in the corresponding sentence.

It was this set of triplets that was used to train and test the encoder-decoder neural network (see below). Note that the number of samples varied across triplets, being determined by how long it took the participant to speak the sentence. The number of channels varied across participants, depending on the number of electrodes implanted (256 [participants **a, b, d**] or 128 [**c**]) or too noisy to be used.

Two sets of sentences were used:

- MOCHA-TIMIT [22]: 460 sentences, $\sim$ 1800 unique words, participants **a, b, d**;
- picture descriptions: 30 sentences, $\sim$ 125 unique words, participants **c, d**.

For both sets, each sentence was presented briefly on a computer screen for recital, followed by a few seconds of rest (blank display). However, to avoid fatiguing participants, no more than 50 sentences were presented in a single session or "block." Thus, MOCHA-TIMIT could not be administered in a single block. To achieve consistency across participants, then, it was first divided into nine subsets, MOCHA-1, MOCHA-2, etc., of 50 sentences apiece (and 60 in MOCHA-9), each of which could be completed within one block, and within which sentence presentations were randomized. This resulted in better coverage of

- MOCHA-1: 50 sentences, $\sim$ 250 unique words, participants **a, b, d**

than the other subsets, and consequently we focus on decoding from it and the 30 picture descriptions in the main results below.

The encoder-decoder network. The architecture was inspired by recent artificial neural networks for machine translation [19], albeit with significant modifications. Abstractly, the encoder module first "consumes" an entire sequence of ECoG data (corresponding to one sentence), which it summarizes in a high-dimensional vector of fixed length, i.e. independent of the number of samples in the input sequence. The encoder then passes this summary to the decoder module, which unpacks it one word at a time.

We now describe our architecture in more detail, following Fig. 3 throughout:

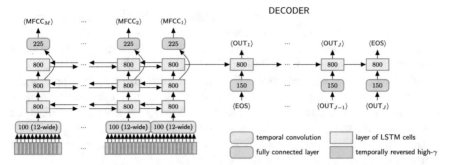

Fig. 3 Network architecture. The network consists of a pair of coupled RNNs, shown here "unrolled" in sequence steps. The encoder consumes ECoG and predicts audio (MFCCs); the decoder is initialized at the final encoder state and predicts words until it emits an end-of-sequence token ($\langle$EOS$\rangle$)

1. *Temporal convolution*: The ECoG "signature" of a particular phoneme or word will not depend on the absolute time at which it was produced: it is time-invariant. Neural networks can efficiently exploit this invariance if they are required to apply the *same* filters at regular temporal intervals ("strides") along the entire length of the input sequence ("temporal convolution"). Furthermore, our filters "strode" by 12 samples, thereby downsampling their inputs from 200 to about 16 Hz. This helps because recurrent neural networks struggle with long sequences [6]. We used 100 filters (each spanning all of the input channels), so the output of the temporal convolutions is a sequence of length-100 vectors.

2. *Encoder recurrent neural network* (RNN): The vectors in this sequence are consumed one at a time (a) in forward order and (b) in backward order by a pair of RNNs, each with 400 units of long short-term memory (LSTM). Then, at every time point, the hidden states of these RNNs are concatenated together, creating a sequence of length-800 vectors. These are the input to a *second* pair of LSTM-based RNNs, which in turn produces inputs for a third pair.

   The hidden states of the second-layer RNNs are also used to predict the sequence of MFCCs, i.e. the speech audio. The hidden state of the deepest (third) RNNs *at the time step of the final sequence element* is interpreted to be a high-dimensional, length-independent summary of the entire input sequence, and is passed to the decoder RNN.

3. *Decoder RNN*: A single-layer, unidirectional, 800-unit, LSTM RNN is initialized at this high-dimensional summary. At each time step it *emits* a probability distribution over all the words in the vocabulary; and *consumes* either the previous word in the sequence (during training) or the previous *most probable word* (during testing). Notice that words, which are encoded as one-hot vectors, are first "embedded" into a dense, 150-dimensional space before entering the decoder RNN.

We make a few technical notes:

- **Time reversal**. Presumably, the first elements of the ECoG data are most related to the first word or words of the sentence. To reduce the number of computation steps separating these, the sequences of ECoG data were temporally reversed before entering the network, following Sutskever and colleagues [19].
- **MFCC sequences**. The sequences of MFCCs therefore also need to be temporally reversed. But they also need to be downsampled, to match the downsampling effect of the strided temporal convolution (see above). We simply decimated the sequences, selecting every twelfth vector (without bothering first to low-pass filter). The purpose of targetting speech audio is simply to guide training onto the right track [5, 20]; during testing, the predicted MFCCs are not used.
- **Training and testing**. The entire network was trained to map ECoG to audio (MFCCs) and text (word sequences) with stochastic gradient descent via back-propagation (with AdaM optimization [10]). Dropout [17] was applied to all layers except the recurrent connections. The remaining details of the training and testing procedure, including hyperparameter optimization, cross-validation, and transfer learning can be found in the original report [11].

3 Results

Decoder performance. Encoder-decoder performance on data from all four participants is shown in the first "violin" of each of the violin plots in Fig. 4. Subfigure labels correspond to participant IDs. Note that participants **a** and **b** read the 50 sentences from MOCHA-1, whereas participants **c** and **d** read the 30 picture descriptions (see **Methods**). The most impressive results are for participant **b**, for whom the encoder-decoder usually achieved WERs close to 0—perfect decoding. Only for participant **a**, who provided only two repeats of each sentence, were WERs outside the acceptable range of speech transcription (25% [15]). For two participants WERs were close to or below 5%, the performance of professional transcribers for spoken speech [23]—albeit with much larger vocabularies.

To understand better the high performance of the encoder-decoder, we trained new sets of networks with certain critical aspects of the architecture or the data removed (Fig. 4):

- **Grid density** ("low density"): Data from a lower-density grid can be simulated by dropping every other electrode from the data. This pseudo-grid will have 8-mm (rather than 4-mm) inter-electrode spacing and one quarter the number of electrodes. Moving to such a grid typically increases (median) WER by about 20 percentage points.
- **Speech audio** ("no MFCCs"): It will be impossible to acquire speech audio from non-speaking subjects—the ultimate target population for a speech prothesis. Training a network without requiring the encoder to predict MFCCs typically increases WER by 15–30 percentage points.

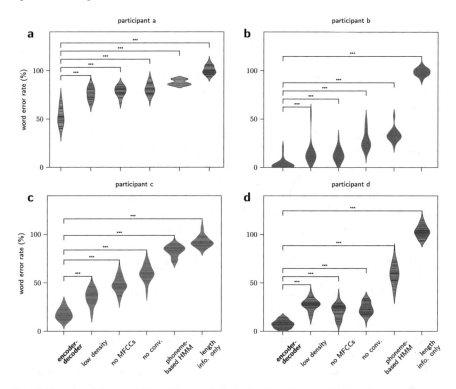

Fig. 4 Decoder performance. For all four participants (**a–d**) and for encoder-decoders as well as various "competitors," text output was evaluated in terms of word error rate (WER), with the sentences actually spoken serving as ground truth. For any single participant and decoder type, the distribution of WERs is across 30 instances of that decoder trained de novo and evaluated on randomly selected held-out blocks. For every participant, the distribution of WERs under the encoder-decoder is significantly better than any competitor ($p < 0.0005$ under a one-sided Wilcoxon signed-rank test, Holm-Bonferroni corrected for five comparisons). In addition to its standard implementation, the encoder-decoder was evaluated under a simulated lower-density grid ("low density"), without audio data during training ("no MFCCs"), without temporal convolution ("no conv."), and with input sequences of pure noise but of the correct length ("length info. only")

- **Temporal convolution** ("no conv."): Using a fully connected input layer amounts to dropping (a) the assumption of time-invariance of the ECoG data, as well as (b) the downsampling. It increased WERs by 20–40 percentage points.

Is the encoder-decoder really just a sentence classifier? For word-based decoding to stand any chance of succeeding, it is necessary to guarantee that the words read during a block used for testing have also been read at least once across the blocks used for training. Given our time constraints, we therefore decided to use, for each participant, a single set of sentences across all training and testing blocks. But this raises the possibility that the encoder RNN is merely *classifying* its inputs—say, with a label from the integers 1–50, which it then hands off to the decoder RNN. The latter could in turn learn how to unpack the 50 labels into their corresponding sentences. This would render the results in Fig. 4 much less general and (therefore) interesting.

The short answer is that the encoder-decoder is *not* merely acting as a sentence classifier. We return to this point in the **Discussion**. Here we show that it performs better than other sentence classifiers, and that *performance improves when it is trained on sentences outside the test set.*

- **Sequence-length information** ("length-info only"): Recall that the ECoG sequences were manually extracted at the sentence boundaries (Fig. 2a). Therefore, if time of production varied more across than within sentence types, it would theoretically be possible for the network to *classify* input sentences based only on their length. We tested this by replacing each ECoG sequence with a sequence of pure noise—but still of the true length—and then re-training and testing networks. This resulted in WERs near 100% for all participant (Fig. 4).
- **Sentence classification** ("phoneme-based HMM"): We compared against a state-of-the-art, HMM-based sentence classifier for neural data [13]. It attempts to decode phonemes from ECoG data and then checks which of the sentences in a closed set (in our case, either MOCHA-1 or the picture descriptions) is most consistent with this phoneme sequence. Using this decoder increased WERs by 30–70 percentage points.
- **Training on non-test-sentences.** For some participants, we were able to collect blocks with sentences outside the test set, in particular MOCHA-2–MOCHA-9. Adding these blocks to a training set originally consisting of two blocks of MOCHA-1 ("+task TL") improved decoding performance on MOCHA-1 by as

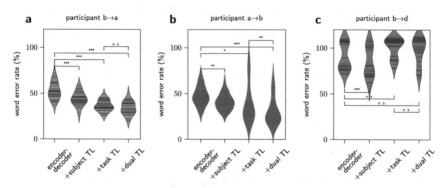

Fig. 5 Performance of the encoder-decoder under *transfer learning*. For the three participants (color code as in Fig. 4) with sufficient coverage of MOCHA-1, 30 encoder-decoders were tested on a randomly held-out block of MOCHA-1. They were trained on 2 blocks of MOCHA-1 ("encoder-decoder"), or on those two blocks *plus*: one block of each of MOCHA-2–MOCHA-9 ("+task TL"); another participant's MOCHA-1 blocks ("+subject TL"); or two participants' MOCHA-1–MOCHA-9 blocks. Significance, indicated by stars (*: $p < 0.05$, **: $p < 0.005$, ***: $p < 0.0005$, n.s.: not significant), was computed with a one-sided Wilcoxon signed-rank test, and Holm-Bonferroni corrected for 14 comparisons: the 12 shown here plus cross-subject transfer learning on the picture descriptions, which did not yield significant improvements

much as 15 percentage points (Fig. 5). This would be impossible if the encoder-decoder were merely classifying sentences of MOCHA-1.

Cross-participant transfer learning. In addition to demonstrating that the encoder-decoder is not acting as a sentence classifier, "cross-task" transfer learning shows how to augment our training data, which is of interest given how limited data-collection time is with epilepsy patients. Still, the total is upper bounded by the amount of a time a participant spends in the hospital. This bound could be breached, however, if we could exploit data from *other participants*. It turns out that we can, by pre-training the encoder-decoder on one participant before training it on another, target participant. On the MOCHA-1 sentences, this cross-participant transfer learning shows significant, albeit modest, reduction in WER for all participants (Fig. 5, "+subject TL"). The two forms of transfer learning can also be combined to yield further improvements (Fig. 5, "+dual TL").

4   Discussion

We decoded speech from ECoG data with error rates near zero, but only in the context of some 50 sentences comprising 250 unique words. The rigidity of the trained network can be seen in some of its errors, when it substitutes a whole sentence of MOCHA-1 for another. On the other hand, we showed that the encoder-decoder is not merely classifying input sentences, since training on non-MOCHA-1 sentences improves performance.

It also turns out that if the encoder is trained on ECoG sequences corresponding to single words (rather than single sentences), the correct word can be identified from its final hidden state with accuracies of up to 80% (for participant **b**; unpublished data), at least on the 250 words of MOCHA-1. The result is not fully general because of coarticulation: a word that appears in MOCHA-1 only after some other word may be produced differently in other contexts. But it very strongly suggests that the architecture can generalize to arbitrary sentences composed from a vocabulary of at least 250 words—given the appropriate training corpus.

The larger remaining questions have to do with clinical translation into patients who have lost the ability to speak. There are at least two problems:

1. It will no longer be possible to train the encoder-decoder to predict speech audio, which will hurt performance (see **Results**).
2. Cortical plasticity post-injury (for example) may obscure or eliminate the relevant neural signals.

In fact, the MFCCs can be replaced with phoneme sequences, sometimes with no drop in performance, and it may be possible (although not easy) to estimate these—or, for that matter, the MFCCs—by controlling the timing of the task. And it may be possible to learn much of the decoder from healthy patients via transfer learning (see **Results**), although it remains to be seen how effectively models transfer to non-speaking patients.

Acknowledgements The project was funded by a research contract under Facebook's Sponsored Academic Research Agreement. Data were collected and pre-processed by members of the Chang lab, some (MOCHA-TIMIT) under NIH grant U01 NS098971. Some neural networks were trained using GPUs generously donated by the Nvidia Corporation.

References

1. Angrick M, Herff C, Mugler E, Tate MC, Slutzky MW, Krusienski DJ, Schultz T (2019) Speech synthesis from ECoG using densely connected 3D convolutional neural networks. J Neural Eng
2. Anumanchipalli GK, Chartier J, Chang EF (2019) Speech synthesis from neural decoding of spoken sentences. Nature 568(7753):493–498
3. Brumberg JS, Kennedy PR, Guenther FH (2009) Artificial speech synthesizer control by brain-computer interface. In: Interspeech, pp 636–639
4. Brumberg JS, Wright EJ, Andreasen DS, Guenther FH, Kennedy PR (2011) Classification of intended phoneme production from chronic intracortical microelectrode recordings in speech-motor cortex. Front Neuroeng 5:1–12
5. Caruana R (1997) Multi-task learning. Multitask Learn 28:41–75
6. Cho K, Van Merrienboer B, Bahdanau D, Bengio Y (2014) On the properties of neural machine translation: encoder–decoder approaches. In: Proceedings of SSST-8, eighth workshop on syntax, semantics and structure in statistical translation, pp 103–111
7. Cho K, Van Merrienboer B, Gulcehre C, Bahdanau D, Bougares F, Schwenk H, Bengio Y (2014) Learning phrase representations using RNN encoder-decoder for statistical machine translation. In: 2014 conference on empirical methods in natural language processing (EMNLP), pp 1724–1734
8. Gehring J, Auli M, Grangier D, Yarats D, Dauphin YN (2017) Convolutional sequence to sequence learning. In: 34th international conference on machine learning, ICML 2017, vol 3, pp 2029–2042
9. Herff C, Heger D, De Pesters A, Telaar D, Brunner P, Schalk G, Schultz T (2015) Brain-to-text: decoding spoken phrases from phone representations in the brain. Front Neurosci 9:1–11
10. Kingma DP, Ba J (2014) Adam: a method for stochastic optimization
11. Makin JG, Moses DA, Chang EF (2020) Machine translation of cortical activity to text with an encoder-decoder framework. Nat Neurosci 23:575–582
12. Martin S, Brunner P, Holdgraf C, Heinze HJ, Crone NE, Rieger J, Schalk G, Knight RT, Pasley BN (2014) Decoding spectrotemporal features of overt and covert speech from the human cortex. Front Neuroeng 7:1–15
13. Moses DA, Leonard MK, Makin JG, Chang EF (2019) Real-time decoding of question-and-answer speech dialogue using human cortical activity. Nat Commun 10(1)
14. Mugler EM, Tate MC, Livescu K, Templer JW, Goldrick MA, Slutzky MW (2018) Differential representation of articulatory gestures and phonemes in precentral and inferior frontal gyri. J Neurosci 4653(46):1206–1218
15. Munteanu C, Penn G, Baecker R, Toms E, James D (2006) Measuring the acceptable word error rate of machine-generated webcast transcripts. In: Interspeech, pp 157–160
16. Pei X, Barbour DL, Leuthardt EC (2011) Decoding vowels and consonants in spoken and imagined words using electrocorticographic signals in humans. J Neural Eng 8(4):1–11
17. Srivastava N, Hinton G, Krizhevsky A, Sutskever I, Salakhutdinov R (2014) Dropout: a simple way to prevent neural networks from overfitting. J Mach Learn Res 15:1929–1958
18. Stavisky SD, Rezaii P, Willett FR, Hochberg LR, Shenoy KV, Henderson JM (2018) Decoding speech from intracortical multielectrode arrays in dorsal "arm/hand areas" of human motor cortex. In: Proceedings of the annual international conference of the IEEE Engineering in Medicine and Biology Society, EMBS, pp 93–97

19. Sutskever I, Vinyals O, Le QV (2014) Sequence to sequence learning with neural networks. In: Advances in neural information processing systems 27: proceedings of the 2014 conference, pp 1–9

20. Szegedy C, Liu W, Jia Y, Sermanet P, Reed S, Anguelov D, Erhan D, Vanhoucke V, Rabinovich A (2015) Going deeper with convolutions. In: 2015 IEEE conference on computer vision and pattern recognition (CVPR). IEEE, pp 1–9

21. Vaswani A, Shazeer N, Parmar N, Uszkoreit J, Jones L, Gomez AN, Kaiser L, Polosukhin I (2017) Attention is all you need. In: Advances in neural information processing systems, pp 5998–6008

22. Wrench A (2019) MOCHA-TIMIT. Online database

23. Xiong W, Droppo J, Huang X, Seide F, Seltzer ML, Stolcke A, Yu D, Zweig G (2017) Toward human parity in conversational speech recognition. IEEE/ACM Trans Audio Speech Lang Process 25(12):2410–2423

Optically Pumped Magnetometers for Practical MEG-Based Brain-Computer Interfacing

Benjamin Wittevrongel, Niall Holmes, Elena Boto, Ryan Hill, Molly Rea, Arno Libert, Elvira Khachatryan, Richard Bowtell, Matthew J. Brookes, and Marc M. Van Hulle

Abstract Brain-computer interfaces (BCIs) analyse neural signatures to decode the user's intention and control an external device. In support of a wide applicability, a reliable non-invasive tool for capturing neural signals with high information content is needed. Currently, the most prominent non-invasive technique is scalp-recorded electroencephalography (scalp-EEG). However, despite being cost-effective and delivering promising results, its limited spatial resolution hampers access to more sophisticated BCI applications. Magnetoencephalography (MEG) might be a better alternative, but is currently vastly underrepresented in the literature as costly and confining acquisition hardware hampers its adoption. Recently, a new generation MEG sensor based on optically pumped magnetometers (OPMs) has been introduced and shown to overcome many of the practical limitations of traditional MEG hardware. However, it is currently unclear whether the OPM-recorded signals are sufficiently stable when used in a BCI context. In this work, we report on a real-time OPM-MEG-based 'mind-spelling' BCI, with which three participants were able to spell words with an average accuracy of 97.7%. This demonstration confirms that single-trial neural responses can be reliably decoded from OPM-MEG and demonstrates its potential for the development of practical MEG-based BCI applications.

Keywords Brain-computer interface (BCI) · Magnetoencephalography (MEG) · Optically pumped magnetometers (OPM) · Steady-state visual evoked potential (SSVEP)

B. Wittevrongel (✉) · A. Libert · E. Khachatryan · M. M. Van Hulle
Laboratory for Neuro- and Psychophysiology, Department of Neurosciences, KU Leuven, Leuven, Belgium
e-mail: benjamin.wittevrongel@kuleuven.be

N. Holmes · E. Boto · R. Hill · M. Rea · R. Bowtell · M. J. Brookes
School of Physics and Astronomy, Sir Peter Mansfield Imaging Centre, University of Nottingham, Nottingham, UK

B. Wittevrongel · M. M. Van Hulle
Leuven Institute for Artificial Intelligence (Leuven.AI), Leuven, Belgium

B. Wittevrongel · A. Libert · E. Khachatryan · M. M. Van Hulle
Leuven Brain Institute (LBI), Leuven, Belgium

© The Author(s), under exclusive license to Springer Nature Switzerland AG 2021
C. Guger et al. (eds.), *Brain-Computer Interface Research*,
SpringerBriefs in Electrical and Computer Engineering,
https://doi.org/10.1007/978-3-030-79287-9_4

1 Introduction

While impressive BCI demonstrations have been reported with invasive neural implants (e.g. electrocorticography [31], depth probes [15], microelectrode arrays [25] or flexible electrode threads [24]), the required surgical intervention cannot be justified for BCI applications that serve a short-term clinical use, such as BCI-assisted neurorehabilitation [28] or cognitive therapies [21], or for those that are aimed at a wider audience, such as neuromarketing [37] or cognitive biometrics for authentication [2, 27]. For these applications, a tool for reliable non-invasive neural recording is preferred.

The vast majority of non-invasive BCI work has focused on scalp-recorded electroencephalography (EEG) [19], using electrodes that are attached to the scalp. While this is a cost-effective and practical recording modality, the spatial precision of scalp-EEG is not optimal, as poor conductivity of skull disperses cortical potentials over a large scalp area where they are mixed with potentials originating from other cortical sources. Magnetoencephalography (MEG) is complementary to scalp-EEG in the sense that it relies on minute variations in the magnetic field induced by electrical neural potentials. MEG does not suffer from spatial blurring and allows researchers to obtain neural signals with a higher spatial precision [6]. However, the adoption of MEG for the development of BCI applications is currently hampered by the expensive and impractical recording hardware based on superconducting quantum interference devices (SQUIDs). SQUID-based MEG requires constant cryogenic cooling and severely confines the participant's movements as the recordings are highly susceptible to movement artefacts [22]. Furthermore, as the MEG helmet is optimised for adults, the recruitment of children for MEG-studies is not trivial.

In this study, we examine a promising novel technique for MEG-BCI based on optically pumped magnetometers (OPMs) [30]. Unlike SQUID sensors, OPMs are small and lightweight sensors, and as they do not require any external thermal regulation [4], the sensors can be placed in contact with the scalp [13] at any location, which is beneficial for the signal-to-noise ratio and the spatial resolution [3]. Previous studies have shown that these sensors can record reliable neural signals from moving individuals [5] and across all age ranges [8].

Despite the promising reports, the OPM technology has not yet been tested in the context of BCI, and it is unclear whether it permits stable single-trial neural responses, as is typically required by more advanced BCI applications. To investigate this, we developed a real-time 'mind-spelling' application based on OPM-MEG which our participants used to spell words by successively gazing at individual characters. We show that single-trial neural responses are robustly captured by OPM-MEG and that this new technology is a viable tool for the development of practical MEG-based BCI applications.

2 Methods

2.1 OPM-MEG System

The experiment was performed using a whole-head multi-channel OPM-MEG system containing 48 second-generation, zero-field magnetometers manufactured by QuSpin Inc. (Colorado, USA). Each sensor is a self-contained unit, of dimensions $12.4 \times 16.6 \times 24.4$ mm^3, containing a Rb-87 gas vapour within a heated glass cell, a laser for optical pumping, and on-board electromagnetic coils for controlling the local magnetic field within the cell. Precisely how this device measures magnetic field has been dealt with in previous reports [30] and this information will not be repeated here. The OPMs were mounted on the participant's head using a rigid, 3D-printed helmet [9] and each sensor was connected via a 60 cm lightweight (3.3 g/m) flex cable to a backpack. Thicker cables were then taken from the backpack to the control electronics. Analogue output signals were fed from the OPM electronics to a National Instruments digital acquisition system (DAQ). Although OPMs can measure two orthogonal components of the magnetic field, we only measured the component of the magnetic field that was normal to the scalp surface in the experiment reported here. Importantly, prior to the start of the experiment, all OPMs were calibrated using a manufacturer established procedure. In brief, on-board-sensor coils were energised to produce a known field within the cell, the output of the sensor was then measured and calibrated to ensure a response of 2.7 V/T.

The system was operated within a magnetically-shielded room (MSR) designed and built specifically for OPM operation (MuRoom, Magnetic Shields Limited, Kent, UK) at the university of Nottingham. This MSR, which comprises two mu-metal layers and a single copper layer, was equipped with degaussing coils [1], effectively reducing the background static magnetic field to ~ 1.5 nT, with field gradients of less than 2 nT/m. The operational dynamic range of the QuSpin zero-field magnetometers (which we define here as the maximum change in field before gain errors become > 5%) is ~ 1.5 nT [5]. In an MSR with a background field of 30 nT, this would mean a head rotation of around 3° is enough to generate a 1.5 nT field change, which would, in turn, cause a significant (> 5%) change in gain of the OPM. In our MSR, an OPM can be rotated through 360° about any axis and still maintain gain error within 5%. Even though OPMs remain operational in the low background field inside our MSR, head movement within this field still generates artefactual signals which can distort measured brain activity. For this reason, the background field and gradients were further controlled using a set of bi-planar coils placed on either side of the participant [11, 12]. These coils, which are wound on two 1.6 m square planes separated by a 1.5 m gap in which the participant is placed, generate three orthogonal magnetic fields and four of the five independent linear gradients within a (hypothetical) 40 cm cube inside which the participant's head is positioned. A reference array, placed behind the participant, then measures the background field/gradient and currents are applied to the bi-planar coils to cancel this remnant field. This takes the background

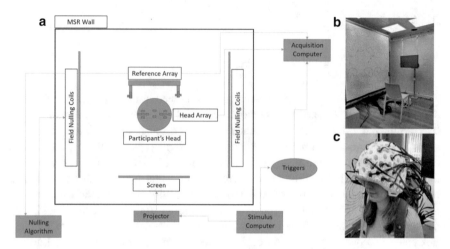

Fig. 1 **a** Schematic diagram of the full OPM setup. Note that during the real-time spelling experiment, the stimulus and acquisition computer were the same device. **b** View inside the magnetically shielded room. **c** Example of the rigid helmet with OPM sensors at different scalp locations

field from 1.5 to ~ 0.5 nT, which enables a three-fold improvement in suppression of movement artefacts.

A schematic diagram of the system is shown in Fig. 1. The participants sat on a non-magnetic chair placed in the centre of the MSR between the bi-planar coils. Note that all control electronics are kept outside the MSR in order to minimise the effect of magnetic interference on the MEG measurements.

2.2 Experimental Paradigm

Subjects Three subjects (1 female, aged 40, 22 and 26 years old, all right-handed) with normal visual acuity participated in the experiment. Subjects were seated in the magnetically shielded room at approximately 80 cm from the projection screen. 48 OPMs were placed in a specially designed helmet [9] and were uniformly distributed across the scalp. The active nulling was dynamic and adapted to small changes in the magnetic field experienced by the OPMs throughout the entire experimental session by using the sensitive outputs of three of the four reference magnetometers (one measurement for each Cartesian component of the magnetic field) as inputs to a high-speed proportional integral controller [12].

All participants provided written informed consent for all experiments. All experiments were approved by the University of Nottingham Medical School Ethics Committee.

Interface The experimental interface consisted of nine squares in a 3 × 3 matrix design displayed on a projection screen using an GT1080 Darbee projector (Optoma,

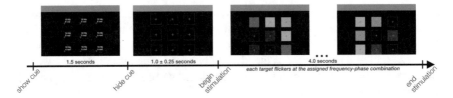

Fig. 2 Visual rendition of one trial in the training session. Note that the frequency-phase combinations shown in the first panel are for exposition purposes and were not shown during the actual experiment

UK) operating at a refresh rate of 60 Hz. Each square spanned a visual angle of 4.3° in the horizontal and vertical dimensions. The inter-square distance was 2.7°, horizontally and 1.7° vertically. Each of the nine squares was assigned a unique frequency-phase combination, as shown in Fig. 2.

Training session Prior to the real-time spelling session, the participants first completed a training session aimed at collecting data for training the classifier. During this session, each square was overlaid with a fixation cross that spanned a visual angle of 0.86°. A training trial started with a visual cue during which one of the nine fixation crosses adopted a red color, and the subject was asked to redirect his/her gaze to this target. After a jittered interval of 1.0 ± 0.25 s, the cue was removed and the nine targets started flickering at their assigned frequency-phase combinations, achieved by adopting a sinusoidal luminosity profile [20]. After 4 s, the flickering stopped and the trial ended. Each target was cued 8 times in pseudorandom order (block design), leading to a total of 72 four-second trials. The total training session lasted approximately 8 min. Data was collected continuously throughout the duration of the training session at a sample rate of 1200 Hz.

Data processing The raw OPM data collected during the training sessions was filtered between 4 and 40 Hz using a fourth-order zero-phase Butterworth filter, cut into four-second epochs locked to the onset of each trial, downsampled to 150 Hz, and labeled with the frequency-phase combination of the corresponding gazed target.

Decoder From the preprocessed data, a classification pipeline based on spatiotemporal beamforming was trained. For each of the nine unique frequency-phase combinations, a spatiotemporal beamformer [33, 35] was constructed that estimates the contribution of the corresponding frequency-phase combination into the current segment of data. The beamformer for target i ($i \in [1 \ldots 9]$) was calculated by obtaining an activation pattern $a_i \in \mathbb{R}^{1 \times mn}$ and a regularised covariance matrix $\Sigma_i \in \mathbb{R}^{mn \times mn}$, where m is the number of channels and n the number of samples in two periods of the corresponding frequency. First, all epochs during which target i was cued were cut into (50%) overlapping segments with length equal to two periods of the stimulation frequency f_i of target i. The activation pattern was then obtained as the average segment $A_i \in \mathbb{R}^{m \times n}$ and vectorised to obtain $a_i \in \mathbb{R}^{1 \times mn}$. The covariance matrix Σ_i was estimated from all available epochs by extracting segments using the procedure mentioned above. Note that epochs not corresponding to the stimulation frequency under consideration are cut into segments of length n. A regularisation constant of 0.95 was adopted in the calculation of the covariance matrix: $\Sigma = \alpha \Sigma + (1 - \alpha)I$, where I is the identity matrix. Using the constraint $a_i w_i = 1$, the linearly-constrained minimum-variance (LCMV) beamformer for target i can then be calculated as follows [32]:

$$w_i = \frac{a_i \Sigma_i^{-1}}{a_i \Sigma_i^{-1} a_i^{\mathrm{T}}},$$

where Σ_i^{-1} is the pseudo-inverse of Σ_i.

Given an epoch, a prediction was made by iteratively extracting the average segment for each target and applying the corresponding beamformer. The winning frequency-phase combination corresponded to the beamformer with maximal output. For a more detailed description of the classification scheme, please see [34–36].

Channel selection To reduce the dimensionality of the decoding model, a greedy forward channel selection strategy was adopted. Starting with an empty set, every iteration defines candidate sets as the currently selected channels extended with each of the non-selected channels. The channel that increases the decoding performance the most is added to the final set. Candidate channel sets were scored using a four-fold cross-validation on the training epochs. Of the 48 OPM channels recorded during the entire experiment, we only considered the 24 OPM sensors and 45 gradiometers located over the parieto-occipital scalp area to speed up the channel selection procedure. In total, the channel selection procedure lasted about two minutes, during which the subject was asked to relax while waiting for the spelling session to start.

Real-time spelling Following the training session, channel selection and classifier training, the real-time spelling session was initiated. Subjects were asked to spell five predefined words. A block began with the presentation of the word-to-spell listed at the top of the display. The fixation crosses were replaced by the characters that are required to spell the displayed word, distributed in a random fashion, one of which was a backspace icon the subject could use to undo previous selections. In case the word only required six or less different characters, the remaining fixation crosses were replaced with other randomly chosen characters. All characters spanned a visual angle similar to the fixation crosses (Fig. 3). Following a 20-s habituation period, the spelling procedure started. A two-second flickering stimulation was presented and the corresponding brain responses were obtained in real-time from the OPM sensors. The recorded data were then filtered as before and submitted to the classification pipeline. For each of the nine beamformers, the two-second epoch was cut into segments of the corresponding frequency, and the segments corresponding to the initial 150 ms were removed. The remaining segments were averaged and applied to the trained beamformer to obtain an estimate (i.e., score) of the presence of the corresponding frequency-phase combination. The beamformer with the highest score was taken as winner and the corresponding character highlighted in yellow. The progression of the

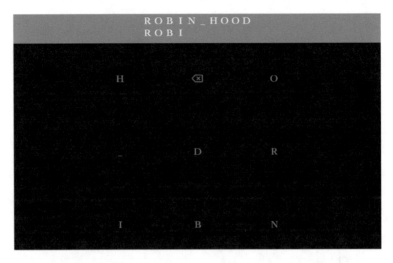

Fig. 3 The interface for the spelling session showed the word to be spelled, as well as the current set of letters that had been sequentially selected by the subject in their attempt to spell out the word. Eight of the crosses have been replaced with the characters required to spell the word, and one cross was replaced with a backspace icon that could be used to undo a previous selection. No cues are given and the stimulation length was reduced to 2 s

character-by-character selection was also displayed at the top of the display, under the word to be spelled. This procedure was repeated until the spelled word contained as many characters as the target word.

Post-hoc analysis While the real-time session was performed using a two-second stimulation, in a post-hoc simulation also shorter stimulation lengths were considered to assess decoding accuracy. To this end, all epochs from the spelling session were shortened by retaining the initial t seconds and presented to the classifier trained on the training set. The predictions were then compared to the actual gazed characters and reported in Fig. 4. This procedure was repeated for increasing signal lengths t from 250 ms to 2 s in steps of 250 ms.

3 Results

Figure 4 shows a typical neural steady-state response to the flickering stimulation. As expected, the SSVEP response is most prominent over the occipital cortex, and unlike a typical scalp-EEG SSVEP in which the response is spatially blurred over a larger scalp area, a more localised response is found with MEG. Additionally, a clear bipolar spatial response profile is obtained as a result of the magnetic properties of a neural dipole.

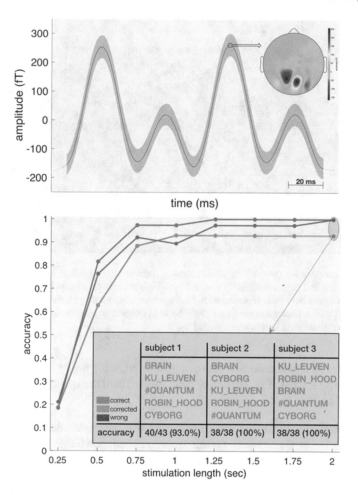

Fig. 4 (top panel) A typical neural steady-state response to flickering stimulation. The full line indicates the average response over a few segments of the same gazed frequency-phase combination and the shaded area the 95% confidence interval. The inset shows the spatial distribution of the response when the amplitude reaches is maximal value. (bottom panel) Real-time spelling results and post-hoc simulation of the decoding accuracies with shorter stimulation lengths. The table shows the performance of the subjects during the real-time mind-spelling session

All three subjects were able to accurately control the spelling application and complete the five words. Using the training data, the greedy channel selection procedure selected nine, seven and four channels. Not unsurprisingly, as our stimuli rely on visual processing, all selected channels were located over the occipital scalp area. The decoder accurately identified the gazed character for all participants with spelling accuracies of 93.02% for the first subject and 100% for the other two subjects (Fig. 4). The three misclassifications that occurred for subject 1 were all successfully corrected by the participant by gazing at the backspace icon, after which the correct character was selected. Accounting for these successful corrections, all three participants correctly spelled all five words.

To investigate the effect of a reduced stimulation length on the decoding accuracy, we ran an offline post-hoc analysis in which we only used the first t (t in [0.250, 2.00] s in steps of 250 ms) seconds of each online trial. The analysis shows that the decoding accuracy is minimally affected by a reduction of the stimulation length until about 750 ms. A stimulation length of 0.25 and 0.5 s elicit a considerable drop in accuracy, similar to what has been described with scalp-EEG [36].

4 Discussion

Non-invasive brain-computer interfacing is mostly accomplished using scalp-recorded EEG, despite its poor spatial resolution due to the inhomogeneous conductivity profile of the head (i.e., cerebrospinal fluid, skull and skin layers). The complementary magnetic field changes are less distorted and allow for neural activity recorded with higher signal-to-noise ratios, but were previously only obtainable using expensive and impractical recording hardware based on SQUIDs, which hampers the development of MEG-based BCIs. Indeed, the number of reports on MEG-based BCIs pales in comparison to their scalp-EEG counterparts, and the majority of these studies target the decoding of (imagined) limb movement [10, 16, 18, 23, 26] or mental tasks [29] as these paradigms allow the participant's head to remain motionless within the MEG helmet. Many other BCI paradigms that are routinely adopted with scalp-EEG have no counterpart in the MEG literature.

In this work, we showed that a new generation of MEG sensor based on OPMs can effectively be deployed for BCI purposes. Our proof-of-concept demonstration shows accurate control of our 'mind-spelling' application and further offline simulations reveal decoding accuracies highly similar to those reported in invasive- and scalp-EEG studies [36]. However, unlike scalp-EEG, which typically requires the application of conductive gel to reduce impedance levels and ensure good quality signals, OPM-MEG does not require the application of additional substances. The helmet containing the OPM sensors can readily be placed on the participant's head without any preparation. OPM-MEG furthermore allows for faster turn-around times: the de-Gaussing and nulling procedure can be completed within minutes and the post-recording cleanup is considerably less labour-intensive given the absence of conductive gel.

Unlike traditional SQUID-based MEG, movement artefacts are less pronounced with OPMs and previous reports have shown reliable OPM-MEG recordings during which the participants were moving [5]. This opens up MEG for new initiatives for BCI-assisted neurorehabilitation programs [14, 17] even beyond hand or arm motions [7] and BCI applications for patients that suffer from involuntary muscular activity, such as in spastic cerebral palsy. Given the high similarity in practical use of OPM-MEG and scalp-EEG, many of the paradigms that are routinely adopted in the latter can readily be adopted using OPM-MEG.

It is worth noting that the goal of the current study was to demonstrate the feasibility of adopting OPM-MEG for BCI research. In this work, we used an SSVEP decoder that was previously developed for decoding scalp-EEG signals [35]. It is very likely that the results presented in this manuscript can be improved by developing algorithms that take advantage of the unique signal properties of the MEG signal. For example, the higher spatial resolution might make is beneficial to consider activations in source space (i.e. source localisation) rather that sensor space. Furthermore, as the OPM technology is still under active development, further improvements to the signal quality and capabilities can be expected.

Conflicts of Interest E.B. and M.J.B. are directors of Cerca Magnetics, a newly established spin-out company whose aim is to commercialise aspects of OPM-MEG technology.

Funding This study was part of a research exchange between KU Leuven and the University of Nottingham for which the lead author was awarded a grant from the Research Foundation—Flanders (V441719N). BW is supported by a post-doctoral mandate from KU Leuven (PDM/19/176). AL is supported by a strategic basic research grant awarded by the Research Foundation—Flanders (1SC3419N). MMVH is supported by research grants received from the European Union's Horizon 2020 research and innovation programme under grant agreement No. 857375, the Financing Program (PFV/10/008) and the special research fund of the KU Leuven (C24/18/098), the Belgian Fund for Scientific Research—Flanders (G088314N, G0A0914N, G0A4118N), the Inter-university Attraction Poles Programme—Belgian Science Policy (IUAP P7/11), and the Hercules Foundation (AKUL 043). Development of the OPM system was supported by the UK Quantum Technology Hub in Sensing and Timing, funded by the Engineering and Physical Sciences Research Council (EPSRC) (EP/T001046/1), and a Wellcome Collaborative Award in Science (203257/Z/16/Z and 203257/B/16/Z).

References

1. Altarev I et al (2015) Minimizing magnetic fields for precision experiments. J Appl Phys 117:233903
2. Barayeu U et al (2020) Robust single-trial EEG-based authentication achieved with a 2-stage classifier. Biosensors 10:124
3. Boto E et al (2016) On the potential of a new generation of magnetometers for MEG: a beamformer simulation study. PLoS One 11:e0157655
4. Boto E et al (2017) A new generation of magnetoencephalography: room temperature measurements using optically-pumped magnetometers. Neuroimage 149:404–414
5. Boto E et al (2018) Moving magnetoencephalography towards real-world applications with a wearable system. Nature 555:657
6. da Silva FL (2013) EEG and MEG: relevance to neuroscience. Neuron 80:1112–1128
7. Foldes ST, Weber DJ, Collinger JL (2015) Meg-based neurofeedback for hand rehabilitation. J Neuroeng Rehabil 12:85
8. Hill RM et al (2019) A tool for functional brain imaging with lifespan compliance. Nat Commun 10:1–11
9. Hill RM et al (2020) Multi-channel whole-head OPM-MEG: helmet design and a comparison with a conventional system. NeuroImage 116995
10. Halme H-L, Parkkonen L (2016) Comparing features for classification of MEG responses to motor imagery. PLoS One 11

11. Holmes N et al (2018) A bi-planar coil system for nulling background magnetic fields in scalp mounted magnetoencephalography. Neuroimage 181:760–774
12. Holmes N et al (2019) Balanced, bi-planar magnetic field and field gradient coils for field compensation in wearable magnetoencephalography. Sci Rep 9:1–15
13. Iivanainen J, Stenroos M, Parkkonen L (2017) Measuring MEG closer to the brain: performance of on-scalp sensor arrays. Neuroimage 147:542–553
14. Jerbi K et al (2011) Inferring hand movement kinematics from MEG, EEG and intracranial EEG: from brain-machine interfaces to motor rehabilitation. IRBM 32:8–18
15. Krusienski D, Shih J (2011) Control of a brain–computer interface using stereotactic depth electrodes in and adjacent to the hippocampus. J Neural Eng 8:025006
16. Lal TN et al (2005) A brain computer interface with online feedback based on magnetoencephalography. In: Proceedings of the 22nd international conference on machine learning, pp 465–472
17. Lazarou I et al (2018) EEG-based brain–computer interfaces for communication and rehabilitation of people with motor impairment: a novel approach of the 21st century. Front Hum Neurosci 12:14
18. Lin PT et al (2013) A high performance MEG based BCI using single trial detection of human movement intention. In: Functional brain mapping and the endeavor to understand the working brain. InTechOpen, pp 17–36
19. Lotte F et al (2018) A review of classification algorithms for EEG-based brain–computer interfaces: a 10 year update. J Neural Eng 15:031005
20. Manyakov NV et al (2013) Sampled sinusoidal stimulation profile and multichannel fuzzy logic classification for monitor-based phase-coded SSVEP brain–computer interfacing. J Neural Eng 10:036011
21. McFarland DJ et al (2017) Therapeutic Applications of BCI Technologies. Brain computer interfaces 47(1–2):37–52
22. Medvedovsky M et al (2007) Artifact and head movement compensation in MEG. Neurol Neurophysiol Neurosci 4
23. Mellinger J et al (2007) An meg-based brain–computer interface (BCI). Neuroimage 36:581–593
24. Musk E et al (2019) An integrated brain-machine interface platform with thousands of channels. J Med Internet Res 21:e16194
25. Pandarinath C et al (2017) High performance communication by people with paralysis using an intra-cortical brain-computer interface. Elife 6:e18554
26. Sabra NI, Wahed MA (2011) The use of meg-based brain computer interface for classification of wrist movements in four different directions. In: 2011 28th national radio science conference (NRSC). IEEE, pp 1–7
27. Rajagopal A et al (2013) Neuropass: a secure neural password based on EEG. Biomed Eng
28. Ramos-Murguialday A et al (2013) Brain-machine interface in chronic stroke rehabilitation: a controlled study. Ann Neurol 74:100–108
29. Spüler M et al (2012) Adaptive SVM-based classification increases performance of a meg-based brain-computer interface (BCI). In: International conference on artificial neural networks. Springer, pp 669–676
30. Tierney TM et al (2019) Optically pumped magnetometers: from quantum origins to multi-channel magnetoencephalography. NeuroImage
31. Vansteensel MJ et al (2016) Fully implanted brain–computer interface in a locked-in patient with ALS. N Engl J Med 375:2060–2066
32. Van Veen BD et al (1997) Localization of brain electrical activity via linearly constrained minimum variance spatial filtering. IEEE Trans Biomed Eng 44:867–880
33. Wittevrongel B, Van Hulle MM (2016) Frequency-and phase encoded SSVEP using spatiotemporal beamforming. PLoS One 11:e0159988
34. Wittevrongel B, Van Hulle MM (2016) Hierarchical online SSVEP spelling achieved with spatiotemporal beamforming. In: 2016 IEEE statistical signal processing workshop (SSP). IEEE, pp 1–5

35. Wittevrongel B, Van Hulle MM (2017) Spatiotemporal beamforming: a transparent and unified decoding approach to synchronous visual brain-computer interfacing. Front Neurosci 11:630
36. Wittevrongel B et al (2018) Decoding steady-state visual evoked potentials from electrocorticography. Front Neuroinform 12:65
37. Yoshioka M et al (2012) Brain signal pattern of engrossed subjects using near infrared spectroscopy (NIRS) and its application to TV commercial evaluation. In: The 2012 international joint conference on neural networks (IJCNN), pp 1–6

EEG Decoding of Pain Perception for a Real-Time Reflex System in Prostheses

Zied Tayeb, Rohit Bose, Andrei Dragomir, Luke E. Osborn,
Nitish V. Thakor, and Gordon Cheng

Abstract *Rationale* In recent times, we have witnessed a push towards restoring sensory perception to upper-limb amputees, which includes the whole spectrum from gentle touch to noxious stimuli. These are essential components for body protection as well as for restoring the sense of embodiment. Despite the considerable advances that have been made in designing suitable sensors and restoring tactile perceptions, pain perception dynamics and how to decode them using effective bio-markers are still not fully understood. *Methods* Here, we used electroencephalography (EEG) recordings to identify and validate a spatio-temporal signature of brain activity during innocuous, moderately more intense, and noxious stimulation of an amputee's phantom limb using transcutaneous nerve stimulation (TENS). *Results* Based on the spatio-temporal EEG features, we developed a system for detecting pain perception and reaction in the brain, which successfully classified three different stimulation conditions with a test accuracy of 94.66%, and we investigated the cortical activity in response to sensory stimuli in these conditions. Our findings suggest that the noxious stimulation activates the pre-motor cortex with the highest activation shown in the

Z. Tayeb (✉) · G. Cheng
Institute for Cognitive Systems, Technical University of Munich, Munchen, Germany
e-mail: zied.tayeb@tum.de

R. Bose · A. Dragomir
N.1 Institute for Health, National University of Singapore, Singapore, Singapore

R. Bose
Department of Bioengineering, University of Pittsburgh, Pittsburgh, USA

A. Dragomir
Department of Biomedical Engineering, University of Houston, Houston, USA

L. E. Osborn · N. V. Thakor
Department of Biomedical Engineering, Johns Hopkins School of Medicine, Baltimore, USA

L. E. Osborn
Research and Exploratory Development, Johns Hopkins University Applied Physics Laboratory, Baltimore, USA

N. V. Thakor
Department of Biomedical Engineering, National University of Singapore, Singapore, Singapore

© The Author(s), under exclusive license to Springer Nature Switzerland AG 2021
C. Guger et al. (eds.), *Brain-Computer Interface Research*,
SpringerBriefs in Electrical and Computer Engineering,
https://doi.org/10.1007/978-3-030-79287-9_5

47

central cortex (Cz electrode) between 450 and 750 ms post-stimulation, whereas the highest activation for the moderately intense stimulation was found in the parietal lobe (P2, P4, and P6 electrodes). Further, we localized the cortical sources and observed early strong activation of the anterior cingulate cortex (ACC) corresponding to the noxious stimulus condition. Moreover, activation of the posterior cingulate cortex (PCC) was observed during the noxious sensation. *Conclusion* Overall, although this is a single case study, this work presents a novel approach and a first attempt to analyze and classify neural activity when restoring sensory perception to amputees, which could chart a route ahead for designing a real-time pain reaction system in upper-limb prostheses.

Keywords Brain computer interface (BCI) · Electroencephalography (EEG) · Noxious stimulation · Spatio-temporal signatures · Reflex system in prostheses

1 Introduction

Nociception is commonly known as the sense of pain [1]. Specialized receptors called nociceptors that cover the skin and organs react to harmful chemical, mechanical and thermal stimuli [2]. Some of these microscopic pain receptors react to all kinds of noxious stimuli, while others only react to specific pain like burning or pricking your finger on something sharp. Jolts of sudden pain activate the A-type fibers to send an electrical signal up to the spinal cord [3]. Pain signals then activate the thalamus, which relays the signal to the different brain regions [4]. Subsequently, the signal activates the somatosensory cortex, which is responsible for physical sensations. The signals are then relayed to the frontal cortex, where higher-order cognitive processing occurs, and finally to the limbic system, which is linked to emotions [5]. This pain processing network, along with pain reflex pathways in the spinal cord [3], are considered of the utmost importance for protecting the body from damaging stimuli [6]. These insights into brain networks have therefore spurred research on unraveling the processes within the body that lead to the unpleasant sensation of pain [7] and on understanding the pain perception mechanism in the brain [8]. Authors in [9] investigated perceptual, motor, and autonomic responses to short noxious heat stimuli using electroencephalography (EEG) and confirmed that pain perception is subserved by a distinct pattern of EEG responses in healthy subjects. Functional magnetic resonance imaging (fMRI) was used in [10] to demonstrate pain-related activation of the anterior cingulate cortex (ACC) and the posterior cingulate cortex (PCC) during transcutaneous electrical nerve stimulation (TENS) in healthy participants. A template of nociceptive brain activity that is sensitive to analgesic administration and suitable for clinical trials and research investigations was identified and validated in [11]. Furthermore, different somatosensory evoked potential (SEP) components and latency differences after stimulation of proximal and distal sites of the median nerves were studied and identified in eight healthy right-handed males [12]. Similarly, other previous studies showed that primary and secondary somatosensory cortices, insular cortex, anterior cingulate cortex (ACC), prefrontal cortex (PFC), and thalamus are activated during experimental pain stimuli [13]. Authors in [14] showed the important role of the parietal lobe in pain perception and understanding. Notwithstanding the enormous number of studies on pain perception and brain responses to different painful stimuli using EEG and fMRI, most of these studies focused on studying brain responses in healthy subjects and have not investigated brain responses when perceiving the sense of pain in amputees nor in human–robot interaction settings [6]. It has, therefore, become imperative to study amputees' brain activity when integrating the sense of touch and pain in their arm prostheses [6]. The core novelty and the main contribution of this paper reside in the use of non-invasive EEG activity to analyze somatosensory evoked responses recorded when receiving three different types of stimulations. These stimulations were chosen to convey different sensation profiles, ranging from pleasant to uncomfortable sensation. For that, we identified a brain activity template during innocuous (INNO), moderately intense (MOD) and noxious (NOX) stimulation of an amputee's phantom hand delivered through a transcutaneous nerve stimulation system (TENS) [15]. Based on the identified spatio-temporal brain activity patterns, we developed an offline system for detecting pain reaction in the brain which can recognize the three stimulation conditions from recorded EEG responses by using effective spatio-temporal bio-markers to identify the different brain regions involved in noxious stimuli processing as well as latency responses for each stimulation condition. The overall goal of this study was to extend upon the work performed by Osborn et.al [6], where the reflex system was implemented in the arm prosthesis and the amputee was not involved in the withdrawing reaction. This is thought to be of the utmost importance when designing a better bidirectional-control system between the human and the prosthesis, and hence increase the amputee's sense of embodiment and the sense of ownership [16]. Additionally, detecting this perceived pain sensation and reaction would have an important role in protecting the prosthesis from being damaged by external stimuli [17]. To the best of our knowledge, even though the presented results are from a case study, this work is among the very few to investigate brain responses to different types of NOX and INNO stimuli and the first study to investigate and characterize spatio-temporal brain activities in amputees during a range of noxious and innocuous sensory feedback to the phantom hand, en route to designing a real-time withdrawal system in upper-limb prostheses. Extending upon the findings of the aforementioned studies, we also investigate attention and perceptual brain circuitry involved in the withdrawal reaction. An overview of the real-time withdrawal system in upper-limb prostheses is shown in Fig. 1. This research was published recently in Nature Scientific Reports journal [18].

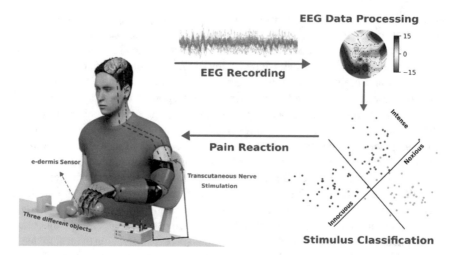

Fig. 1 System implementation overview of a prosthetic arm that can restore the sense of touch and pain

2 Methods

2.1 Patient Recruitment and Sensory Stimulation

An amputee participant (29 years old) with a bilateral amputation more than five-years prior to the current experiments, (due to tissue necrosis from septicemia), was recruited at Johns Hopkins University in Baltimore to perform a series of an embodied prosthesis control as well as sensory feedback experiments. The participant has a transhumeral amputation of the left arm and a transradial amputation of the right arm. All sensory feedback and prosthesis experiments were performed on the participant's left arm. EEG data were collected in one session over a period of two hours. For the sensory stimulation, we performed sensory mapping of the amputee's phantom hand through transcutaneous electrical nerve stimulation (TENS) using a 1-mm beryllium copper (BeCu) probe connected to an isolated current stimulator (DS3, Digitimer Ltd., Hertfordshire, UK). An amplitude of 0.8 mA and frequency of 2 to 4 Hz were used while mapping the phantom hand. The amputee identified areas of phantom activation during sensory mapping and the stimulation sites were noted using anatomical and ink markers. For the stimulation experiment, we used 5-mm disposable Ag–Ag/Cl electrodes on the residual limb sites that mapped to the thumb/pointer, pinky/ulnar, and wrist of the phantom hand. The stimulation sites were the same as those used in [6]. It should be noted that sensory mapping was only performed on the left (transhumeral) residual limb because the amputee participant only wears a prosthesis on his left (transhumeral) side and not his right (transradial) side. This study was carried out in accordance with the Declaration of Helsinki. All experiments were approved by the Johns Hopkins Medicine Institutional Review

Boards. The participant was asked to sign a written informed consent and he agreed to take part in all our experiments.

2.2 EEG Data Recording and Experiment

Brain activity correlates of transcutaneous electrical nerve stimulations were investigated by recording 64-channel EEG data from the amputee participant. Different locations on the participant's left residual limb were identified so that, when stimulated, they activate different regions of the participant's phantom hand. In this EEG experiment, the subject was seated comfortably and was looking at a black cross on a white wall. EEG recordings during various stimulations of the subject's peripheral nerve sites corresponding to the thumb/pointer finger, pinky/ulnar side of the hand, and the wrist of his phantom hand. We stimulated the subject's residual limb in regions that activated his phantom hand using transcutaneous electrical nerve stimulation (TENS). The stimulation included three different conditions for the thumb/pointer and two conditions for the other sites. All values of stimulation were based on previous mapping and psychophysics with this subject [6]. All three conditions (INNO, MOD, NOX) were applied to the thumb/pointer stimulation site and the INNO and MOD conditions were applied to the pinky/ulnar and wrist stimulation sites. Blocks of each stimulation condition were randomly presented as five consecutive stimulation pulse trains lasting for 2 s with a delay of 4 s. Stimulation condition blocks were presented 4 times, yielding a total of 60 trials for the three conditions. A break of 2 min was given between stimulation blocks, and a break of 10 min was given between the different stimulation sites. Condition 3 (NOX) was only presented to the thumb/pointer stimulation site, whereas Conditions 1 and 2 (INNO and MOD) were presented to all stimulation locations (thumb/pointer, pinky/ulnar, and wrist). Condition 3 was only presented to the thumb/pointer location to reduce the total time the subject experienced noxious sensations. EEG data were collected using a 64 channel EEG device (Neuroscan system) with a 500 Hz sampling rate. The montage used the 5% 10/20 system. Electrode impedance was kept below 10 kOhm in at least 95% of derivations throughout the experiment. The amplitude of the transcutaneous electrical nerve stimulation was 1.6 mA for all sites of stimulation and the subject rated each condition's discomfort level using a comfort scale. To ensure that the subject did not substitute or anticipate the stimuli by sight, the EEG data were recorded without the subject wearing the prosthesis.

2.3 EEG Signal Processing and Classification

EEG data were recorded at 500 Hz. The reference electrode was chosen on the vertex and the ground electrode was located on the forehead. Data were processed with specially designed Jupyter notebooks in Python using both gumpy [19] and MNE

[20] toolboxes. For data analysis, 60 trials in total for the three stimulation conditions were used. EEG signals were band-pass–filtered between 0.5 and 70 Hz using a fourth-order Butterworth filter and notch filtered thereafter at 60 Hz. Muscle artifacts were rejected by the Automatic Artifact Rejection (AAR) [21] and independent component analysis (ICA) was used to remove eye movement artifacts [22]. EEG data collected over several trials of the same experiment were averaged together. All EEG scalp topographies were plotted using the MNE toolbox, by matching channel location with its value given the defined latency. Topographies are color encoded, where the green or yellow present null values, the blue color presents negative values, and the red encodes positive values. The color intensity correlates with the channel value. Chosen time latencies in the topographic maps were chosen based on an algorithm [20] that computes and finds the highest peaks at each time point from all electrodes. For feature extraction and classifying the three conditions from EEG, we implemented and tested a wide range of classical machine learning approaches which are based on hand-crafted features. Five different classifiers from the gumpy. Classification module [22] were used and evaluated: K-Nearest Neighbor (KNN), Support vector machine (SVM), Naive Bayes (NB), Linear Discrimination Analysis (LDA), and Quadratic Linear Discrimination Analysis (QLDA). Two different feature extraction methods were used, namely, the maximum amplitude value computed from each channel for a fixed 100 ms time-window, yielding a total number of 64 features (number of electrodes) as well as common spatial patterns (CSP) [23]. The CSP method yielded slightly lower results (a mean accuracy of 85%) than the maximum amplitude value and was therefore discarded in our further analysis. Two different post-processing methods were investigated and tested. First, a principal component analysis (PCA) method with only two components was used for dimensionality reduction, and the two components were fed thereafter to the different classifiers. Second, we further investigated keeping all the extracted 64 features and we used a feature selection algorithm [24] to select the most discriminating subset of features (channels). Data were divided into 80% for training and 20% for testing. Overall, tenfold cross-validation was performed on training data to validate the model (validation accuracy) and the remaining 20% were using for the test phase. For all analyses, balanced accuracy (bACC) was chosen as an evaluation metric for the trained models. bACC is calculated as the average of the proportion corrects of each class individually, where the same number of examples in each class was used. Overall, we wish to mention that the first feature extraction method (max amplitude value) combined with PCA using SVM clearly outperformed the other investigated methods, yielding a validation accuracy of more than 95% and test accuracy of more than 94%. A grid search was performed to select the best hyperparameters of the SVM classifier for a given tenfold cross-validation.

2.4 Source Localization

The MNE toolbox [20] combined with gumpy [19] Python toolbox was used for EEG processing and for source localization. First, we performed cortical surface reconstruction using FreeSurfer [25]. Second, the forward solution and the forward model were computed using the boundary-element model (BEM) [26]. Thereafter, the regularized noise-covariance matrix, which gives information about potential patterns describing uninteresting noise source, was computed and estimated. Afterward, we computed the singular value decomposition (SVD) of the matrix composed of both estimated noise-covariance and the source covariance matrix. Finally, dynamic statistical parametric maps (dSPM) [27] was computed and used for source localization and reconstruction. For dSPM, an anatomical linear estimation approach is applied. This assumes the sources are distributed in the cerebral cortex [27]. A linear collocation single-layer boundary-element method (BEM) [26] is used to compute the forward solution which models the generated signal pattern at each location of the cortical surface. A noise-normalized minimum norm estimate is estimated at each cortical location resulting in an F-distributed estimation of the cortical current. Overall, dSPM identifies the locations of statistically increased current-dipolar strength relative to the noise level.

3 Results

3.1 All Stimulation Conditions Activate the Parietal Lobe. Noxious Stimulation Activates the Central Motor Cortex

Sensory feedback of the three conditions (NOX, MOD, and INNO) tactile stimuli was delivered to the phantom hand using TENS on the transhumeral amputee's residual limb. Our analysis shows that all stimulation conditions elicit early activation of the parietal lobe (around 54 ms) that persists over time for all types of sensation. Interestingly, the MOD stimulation elicits higher activation of the parietal lobe compared to the INNO and the NOX stimulations. In contrast to both INNO and MOD stimulations, only the NOX stimulation activates the central cortex. Based on our findings, we postulate that the NOX stimulation started in the parietal (54 ms) and centro-parietal lobe and rapidly activated the perceptual mechanism in the subject's brain, but then moved towards the pre-motor and central cortex, which could explain that the NOX stimulation activated the pain perception and reaction mechanism in the brain.

3.2   Spatio-Temporal Biomarkers for Noxious-Evoked Activity

We then extended this analysis by seeking to identify a spatio-temporal template to distinguish between the three conditions and find the exact brain response time and spatial location. For the NOX stimulation, the highest activation was found at the central cortex (Cz) in the post-stimulation time window from 450 to 750 ms when comparing it to the INNO stimulation, and the EEG background activity. In contrast with the NOX stimulation that shows high activation of the central cortex, the MOD stimulation was found to be high at the P2, P4, and P6 electrodes for the whole second of analysis (Fig. 2).

3.3   Successful Classification of the Three Different Stimulation Conditions

Using specific spatio-temporal biomarkers for the classification of the three different stimulation conditions, a classification accuracy of more than 94% was achieved in the test phase (Fig. 3).

3.4   Noxious-Related Activation Within the Medial Wall of the Cerebral Cortex

By analyzing the EEG activity at the source level, we found using that NOX sensation elicits activation of the centro-parietal lobe, activation of the anterior cingulate cortex (ACC), the somatosensory motor cortex, and the posterior cingulate cortex (PCC). Overall, the activation of the ACC presents direct evidence that it plays a role in activating the attention circuitry in the brain as well as an important role in external sensory stimuli perception. Moreover, our study reveals that the NOX stimulation activates the PCC (Fig. 4).

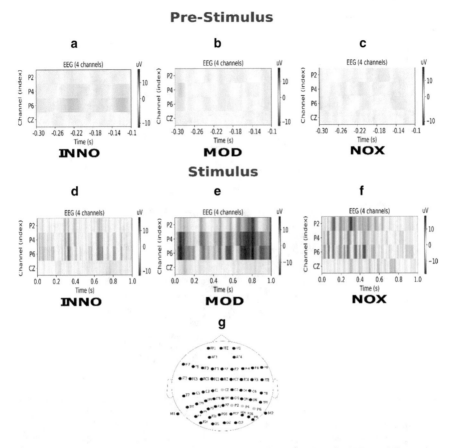

Fig. 2 EEG activity for the parietal and central cortex electrodes. Panels A, B, and C represent EEG activity in Cz, P2, P4, and P6 during the pre-stimulus phase during the INNO, MOD, and NOX stimulation, respectively. Panels D, E, and F represent EEG activity in four different electrodes: Cz (middle cortex) and P2, P4, and P6 electrodes in the parietal lobe during the INNO, MOD, and NOX stimulation, respectively. When comparing panel D to E, a parietal enhancement (red color) and a central depression (blue color) are observable. When comparing panel D and E to F, a central enhancement is observable (red color). Panel G shows the sites used in this study based on the 10–20 system, with the Cz, P2, P4, and P6 electrode's positions highlighted in blue

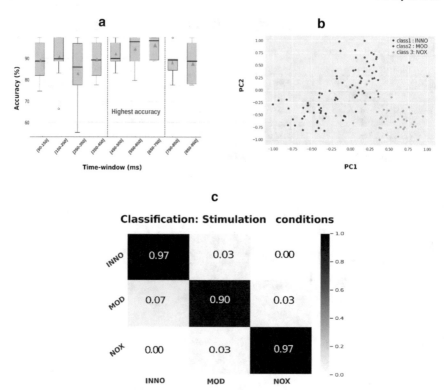

Fig. 3 Classification results of the three stimulation conditions. Panel A is the validation accuracy in different time-windows between 50 and 1000 ms after stimulation represented in a boxplot, showing that the highest validation accuracy was obtained in the time-window 650–750 ms. The green triangle represents the mean accuracy value for each time window, whereas the black line represents the median value for the same time window. Panel B shows the 2D feature space after performing PCA, highlighting a clear separation between the three conditions. PC1 and PC2 represent the first two components after performing PCA. Panel C reflects the confusion matrix in the time-window with the highest accuracy (650–750 ms, shown in Panel A) when classifying the three stimulation conditions in the test phase

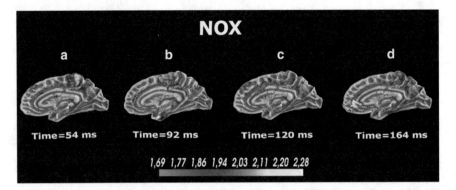

Fig. 4 EEG analysis at the source level for the noxiously evoked activity in the first 200 ms. The dynamic statistical parametric maps (dSPM) [27] was used to compute the reconstructed sources. The scale represents the EEG amplitude activity in uV. Panel A presents high EEG activity in the centro-parietal lobe after 54 ms of stimulation. Panel B shows high EEG activity in the central cortex after 92 ms. Panel C reflects activation of the PCC after 120 ms. Panel shows activation shows activation of the ACC and the parietal lobe after 164 ms

Acknowledgements The authors would like to thank Stefan Ehrlich, Dr. Emmanuel Dean, Nicolas Berberich, and Constantin Uhde for the fruitful discussion. We would also like to thank the Statistical Consulting Service at the Technical University of Munich (TUM) for consultation on our statistical analysis and results. This work was supported in part by Ph.D. grant of the German Academic Exchange Service (DAAD).

References

1. Smith ESJ, Lewin GR (2009) Nociceptors: a phylogenetic view. J Comp Physiol A 195:1089–1106
2. Dubin AE, Patapoutian A (2010) Nociceptors: the sensors of the pain pathway. J clinical investigation 120
3. Skljarevski V, Ramadan NM (2002) The nociceptive flexion reflex in humans—review article. Pain 96:3–8
4. Aziz CA, Ahmad AH (2006) The role of the thalamus in modulating pain. Malays J Med Sci 13:11–18
5. Rolls ET (2013) Limbic systems for emotion and for memory, but no single limbic system. Cortex 62
6. Osborn LE et al (2018) Prosthesis with neuromorphic multilayered e-dermis perceives touch and pain. Sci. Robotics 3
7. Steeds CE (2016) The anatomy and physiology of pain. Surg (Oxford) 34
8. Perl ER (1968) Myelinated afferent fibres innervating the primate skin and their response to noxious stimuli. J Physiol 197
9. Tiemann L et al (2018) Distinct patterns of brain activity mediate perceptual and motor and autonomic responses to noxious stimuli. Nat Commun 9
10. Kwan CL, Mikulis DJ, Davis KD, Crawley AP (2000) Crawley. An fMRI study of the anterior cingulate cortex and surrounding medial wall activations evoked by noxious cutaneous heat and cold stimuli. Pain 85

11. Hartley C et al (2017) Nociceptive brain activity as a measure of analgesic efficacy in infants. Sci Transl Medicine 9
12. Hada Y (2006) Latency differences of N20, P40/N60, P100/N140 SEP components after stimulation of proximal and distal sites of the median nerve. Clin EEG Neurosci 37
13. Ong WY, StohlerDeron CS, Deron, RH (2019) Role of the prefrontal cortex in pain processing. Mol Neurobiol 2
14. Benuzzi F, Lui F, Duzzi D, Nichelli PF, Porro CA (2008) Does it look painful or disgusting? ask your parietal and cingulate cortex. J Neurosci 28:923–931
15. Osborn L et al (2017) Targeted transcutaneous electrical nerve stimulation for phantom limb sensory feedback. In: 2017 IEEE Biomedical Circuits and Systems Conference (BioCAS), Torino, Italy
16. Gouzien A et al (2017) Reachability and the sense of embodiment in amputees using prostheses. Sci Reports 7
17. Troyk PR, Cogan SF (2005) Sensory Neural Prostheses, 1–48. Springer, US, Boston, MA
18. Tayeb Z, Bose R, Dragomir A, Osborn LE, Thakor NV, Cheng G (2020) Decoding of pain perception using EEG Signals for a Real-Time Reflex System in prostheses: a case Study. Sci Rep 10(1):1–11
19. Tayeb Z et al (2018) Gumpy: a Python toolbox suitable for hybrid brain–computer interfaces. J Neural Eng 15:065003
20. Gramfort A et al (2013) MEG and EEG data analysis with MNE-Python. Front Neurosci 7
21. Pion-Tonachini L, Hsu S, Chang C, Jung T, Makeig S (2018) Online automatic artifact rejection using the real-time EEG source-mapping toolbox (rest). In: 2018 40th Annual International Conference of the IEEE Engineering in Medicine and Biology Society (EMBC), 106–109
22. Radüntz T, Scouten J, Hochmuth O, Meffert B (2015) EEG artifact elimination by extraction of ICA-component features using image processing algorithms. J Neurosci Methods 243:84–93
23. Grosse-Wentrup M, Buss M (2008) Multiclass common spatial patterns and information theoretic feature extraction. IEEE Transa on Biomed Eng 8:1991–2000
24. Pudil P, Novovicova J, Kittler J (1994) Floating search methods in feature selection. Pattern Recognit Lett 15:1119–1125
25. Fischl et al (2002) hole brain segmentation: Automated labeling of neuroanatomical structures in the human brain. Neuron 97
26. Kybic J, Clerc M, Faugeras O, Keriven R, Papadopoulo T (2006) Generalized head models for MEG/EEG: boundary element method beyond nested volumes. Phys Medicine Biol 51:1333–1346
27. Dale AM et al (2000) Dynamic statistical parametric mapping: Combining fMRI and MEG for high-resolution imaging of cortical activity. Neuron 26:55–67

High-Dimensional (8D) Control
of Complex Effectors Such
as an Exoskeleton

Alexandre Moly

Abstract In this chapter, we interview Dr. Alexandre Moly about their work with a brain-computer interface (BCI) to control an exoskeleton. Dr. Moly describes how their team attained eight-dimensional control, which is a significant improvement over typical BCIs. High-dimensional control is important for complex tasks such as grasping, which could not only provide a replacement for lost functions but also support rehabilitation. The interview ends with future directions and advice for newcomers to BCI research.

Keywords Brain Computer interface · BCI · ECoG · Clinical trial · Asynchrone · Adaptive · Closed-loop · Online · Brain signal processing · Tetraplegia

1 Introduction

Patients with tetraplegia have paralysis in all four limbs. Many types of exoskeletons could help these patients, but are difficult to control with BCIs or other interfaces. Natural control over limb movements is extremely complicated, and most interfaces allow much less detailed and precise control than the healthy nervous system. BCIs that used ECoG have advanced greatly over the last several years, but we are still a long way from being able to restore control of even one limb that is comparable to natural control, let alone four. As the following quote from Dr. Moly indicates, systems that provide limb control can make a huge difference in some patients' lives.

A. Moly (✉)
CEA, LETI, CLINATEC, MINATEC, University Grenoble Alpes, Grenoble, France
e-mail: alexandre.moly@cea.fr

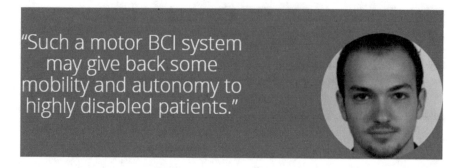

"Such a motor BCI system may give back some mobility and autonomy to highly disabled patients."

Many submissions to the BCI Research Award (and several nominees and winners) have involved ECoG-based contol of exoskeletons. This was the topic of the submission that was one of the second place winners this year. This project entailed a large group of researchers from three French institutions:

High-Dimensional (8D) Control of Complex Effectors Such as an Exoskeleton by a Tetraplegic Subject Using Chronic ECoG Recordings Using Stable and Robust Over Time Adaptive Direct Neural Decoder

Alexandre Moly[1], Thomas Costecalde[1], Félix Martel[1], Antoine Lassauce[1], Serpil Karakas[1], Gael Reganha[1], Alexandre Verney[2], Benoit Milville[2], Guillaume Charvet[1], Stéphan Chabardes[3], Alim Louis Benabid[1], Tetiana Aksenova[1] Alexandre Moly[1], Thomas Costecalde[1], Félix Martel[1], Antoine Lassauce[1], Serpil Karakas[1], Gael Reganha[1], Alexandre Verney[2], Benoit Milville[2], Guillaume Charvet[1], Stéphan Chabardes[3], Alim Louis Benabid[1], Tetiana Aksenova[1]

1 *CEA, LETI, CLINATEC, University Grenoble Alpes, MINATEC, France*
2 *CEA, LIST, DIASI, SRI, Gif-sur-Yvette, France*
3 *Centre Hospitalier Universitaire Grenoble Alpes, France*

This is the first time the BCI Research Awards had a tie for second place. The other project that won second place also involved implanted BCI to help people with movement disabilities (specifically, Parkinson's Disease). To learn more about that project, please see the interview with Tomislav Mikelovic in this book. Here, we present an interview with Dr. Moly about his team's work. You can see more of this project through the video they submitted to the jury.[1] We also wish to congratulate Dr. Moly, who earned his Ph.D. after the 2020 awards while we were developing this book.

---

[1] https://www.youtube.com/watch?v=itCKP8yi1Us.

2 Interview

Hi Alexandre, you submitted your BCI research project titled "High-dimensional (8D) Control of Complex Effectors such as an Exoskeleton by a Tetraplegic Subject Using Chronic ECoG Recordings Using an Adaptive Direct Neural Decoder that is Stable and Robust Over Time" to the BCI Award 2020 and won Second place. Could you briefly describe what this project was about?

Alexandre: Numerous accidents or diseases lead to partial or complete motor disabilities. In France alone, 10,000 spinal cord injuries leading to paraplegia or tetraplegia are registered each year. The "BCI and Tetraplegia" project of CLINATEC (a biomedical research Center including CEA-Grenoble and University Hospital CHU Grenoble-Alpes teams) is a clinical trial in which an online BCI system based on semi-invasive recording allows tetraplegic patients to control complex effectors. We demonstrated that a tetraplegic patient implanted with epidural electrocorticographic (ECoG) implants (WIMAGINE®) could control several limbs from an exoskeleton over a long time without requiring daily recalibration.

What was your goal?

Alexandre: Our primary goal was to provide proof of concept that tetraplegic patients can control complex effectors, such as a 4-limb exoskeleton, thanks to ECoG brain activity monitoring and neural decoding. After the ECoG recording implant surgery, the patient was already able to perform simple task. After 3 years of training, the patient could walk and control both arms of the exoskeleton in real-time. We also showed that the decoding algorithms used for decoding the patient's neural signal provided good and stable decoding performance when updated during several sessions using online incremental closed-loop decoder adaptation procedure. Our next goal is to increase the experiment complexity with tasks such as object grasping to get closer to daily life applications.

What technologies did you use?

Alexandre: This project relied on three innovative key elements: the exoskeleton, the epidural ECoG implant called WIMAGINE®, and the online decoding algorithms used to decode the patient's brain signals.

1. The Enhancing MobilitY (EMY) exoskeleton is a wearable fully motorized four-limb robotic neuroprosthesis weighting 65 kg designed to be driven by the decoded ECoG brain signals.
2. For chronic processing of brain signals, CLINATEC designed an innovative wireless epidural ECoG recording system named Wireless Implantable Multi-channel Acquisition system for Generic Interface with NEurons (WIMAGINE). This fully implantable device records the neural signal at the surface of the cortex above the dura matter before wirelessly sending the signals to the decoding system, which translates the neural signals into commands to the exoskeleton.

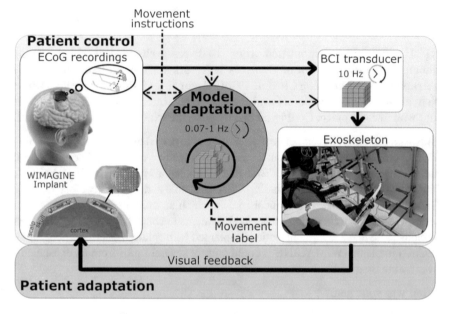

Fig. 1 Clinatec's "BCI and Tetraplegia" clinical trial BCI platform

3. Finally, the decoding algorithms designed for the clinical trial are based on a Mixture of Experts algorithm structure. One regression model is associated with each type of movement of the exoskeleton (such as 3-Dimensional continuous translation of the left hand, 1-Dimensional rotation of the right wrist, etc.). Finally, a Hidden Markov model is calibrated to activate or inhibit the prediction of the regression models to avoid non-zero velocity movements from the non-controlled limb. Each the models are updated in real-time during the online closed-loop experiments in an incremental manner to integrate the patient's visual feedback into the calibration process and therefore allow the model to learn from the patient and the patient from the model (Fig. 1).

What kinds of people could benefit from your research?

Alexandre: Tetraplegic patients may benefit a lot from our research. A motor BCI system like this may restore some mobility and autonomy to highly disabled patients, which is a major concern for social life and professional reintegration. Moreover, the proposed BCI system should also be considered for rehabilitation applications and as a control mechanism for paraplegic patients.

Do you think your work has future potential for clinical use?

Alexandre: Indeed, the results are promising and motivate us for future clinical applications. To our knowledge, this is the first time that an exoskeleton is used by a tetraplegic patient to perform alternative bimanual tasks during long term experiments with a fixed model during more than 6 months. Model stability across time is a major challenge of the BCI field. The proposed solution, which is based on an online incremental closed-loop adaptive decoder, already highlighted numerous benefits to improve the decoding performance and robustness. Our team will achieve further investigation on the patient and model training to improve BCI systems.

How did it feel to win second place in the BCI Award 2020?

Alexandre: CLINATEC is a relatively young laboratory. So, it's really important to us that the BCI community acknowledged our work, especially considering all the tremendous research results that were presented by the other teams. In a more personal point of view, I could not hope for a better way to end my Ph.D.

How can students and researchers get involved in your research?

Alexandre: As previously mentioned, the "BCI and Tetraplegia" clinical trial is based on multiple innovative blocks requiring knowledge from several fields, including: robotics and mechatronics for the exoskeleton; hardware and firmware electronic development for the implant improvement; applied mathematics and signal processing for the neural signal decoding; and neuroscience and medical skills for patients training. Every year, CLINATEC welcomes students to work on the BCI project and many other exciting research topics.

A Computer-Brain Interface that Restores Lost Extremities' Touch and Movement Sensations

G. Valle, F. M. Petrini, P. Mijovic, B. Mijovic, and S. Raspopovic

Abstract *Rationale* Sensory feedback from the lower limbs is essential for correct balance and symmetry during walking. Lower limb amputees suffer from complete lack of sensory feedback from currently available prostheses that exclude the central nervous system from correct sensory-motor integration, causing serious problems. These problems include falls due to unexpected perturbations, asymmetric walking and balance inducing bone pathologies and higher power consumption, and feeling that the prosthesis is a foreign body, with consequent abandonment and phantom limb pain occurrence. Although considerable efforts have focused on developing and controlling sophisticated lower limb prostheses (LLP), few trials have been conducted to restore sensory feedback. *Methods* Three transfemoral amputees have received four intraneural microelectrode arrays in the distal portion of the residual sciatic nerve to electrically stimulate sensations from their missing lower leg and foot. A commercial prosthetic leg (RHEO XC) was equipped with an encoder embedded in the knee and with a custom-made sensorized sole, providing pressure information from 7 locations under the foot sole. The readouts of these sensors and encoder were used to wirelessly drive the stimulation of 4 active sites, eliciting natural touch referred under the foot sole and calf contraction, intuitively interpreted by the subject as knee flexion. We assessed the participants' mobility, confidence, cognitive load, pain level and metabolic cost. We then compared neuroprosthetic control of the bionic leg with that of a commercial-like use of the leg without feedback. *Results* We demonstrated that the natural sensory feedback can be restored in transfemoral amputees. We further showed that these patients can use this feedback to improve their use of the leg prosthesis during different ambulation tasks and promote its integration in their body schema. We designed a neuroprosthetic framework to restore sensory feedback

---

G. Valle (✉) · S. Raspopovic
Laboratory for Neuroengineering, Department of Health Sciences and Technology, Institute for Robotics and Intelligent Systems, ETH Zürich, 8092 Zürich, Switzerland
e-mail: giacomo.valle@hest.ethz.ch

F. M. Petrini
SensArs Neuroprosthetics, C1004 Lausanne, Switzerland

P. Mijovic · B. Mijovic
mBrainTrain D.o.o, 11000 Belgrade, Serbia

C. Guger et al. (eds.), *Brain-Computer Interface Research*,
SpringerBriefs in Electrical and Computer Engineering,
https://doi.org/10.1007/978-3-030-79287-9_7

referred on the phantom lower limb of transfemoral amputees and triggered from the bionic leg by stimulating the residual tibial branch of the sciatic nerve through implanted neural interfaces. *Conclusion* This is the first study of implanted intraneural neuroprostheses for restoring dynamic sensations from the missing leg to people with chronic amputation. It represents a major advance, with a clear translational path, for clinically viable neuroprostheses for restoration of mobility, confidence, metabolic consumption and reduced pain after amputation. These results show that natural invasive sensory feedback restored by means of intraneural electrodes successfully addresses current limitations of prosthetic devices, opening the way for a dramatic improvement of amputees' lives.

Keywords Brain computer interface · BCI · Neuroprosthesis · Neural interfaces · Sensory feedback · Neuromodulation · Intraneural stimulation · Amputees · EEG · Somatosensory

1 Introduction

Lower-limb amputees use commercial prosthetic devices that do not provide proper sensory information to the brain regarding the interaction of the device with the ground or its movement [1]. People with amputation must rely on very limited and uncomfortable haptic information from the stump-socket interaction, and thus face grave impairments. The risk of falls [2], decreased mobility, the perception of the prosthesis as an extraneous body (low embodiment [3, 4]) and the increased cognitive burden during walking with consequent psychological distress and device abandonments [5–7] are some of the most important issues. Costs wasted on an unused prosthetic limb along with a sedentary lifestyle are associated with long-term medical problems (e.g. obesity, diabetes, cardiovascular diseases [7]) and lifelong medical expenses [8].

Even though several research groups have focused on the development and control of sophisticated lower limb prostheses (LLPs) [9, 10], few trials have been dedicated to the restoration of the sensory feedback [11, 12]. In particular, surgical techniques [13] and non-invasive technologies, based on continuous or time-discrete vibrotactile and electro-cutaneous stimulation [14–16], have been adopted to provide the amputees with sensory feedback. The studies mostly reported use in transtibial amputees. These non-invasive sensory feedback devices have demonstrated only limited benefits, such as improved symmetry between prosthetic and healthy legs during walking on even surfaces and postural stability on a movable force platform [14]. Non-invasive technologies have the drawback of not being homologous (the sensation is not perceived as natural and correct one while using the prosthesis) or selective (they evoke defined and spatially-matching sensations) [17]. These drawbacks force the amputees to invest time in training, which only partially overcomes these limitations. Moreover, transtibial amputation is a much less disabling condition than transfemoral amputation. Indeed, transfemoral amputees have less mobility and

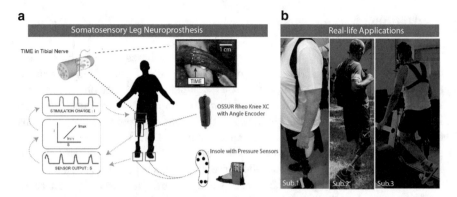

Fig. 1 **a** Somatosensory Neuroprosthesis: the neural implants in the tibial nerve allowed users to interact with the nervous system fibers, providing the patients with artificial sensations. The sensations were linked to the wearable sensor outputs equipping the leg prosthesis. **b** This novel technology was tested in three transfemoral amputees inside and outside the lab

gait symmetry, together with higher energy expenditure than transtibial ones [18, 19]. A novel surgical procedure (agonist–antagonist myoneural interface (AMI) [13, 20]) to restore proprioception in transtibial amputees has been developed. Notably, performance characterization of this approach in daily life activities was not shown yet, and this procedure might be difficult to transfer to higher-level amputations.

After an amputation, the neural pathways between the remaining periphery and the brain are still functional. Thanks to the use of implantable neural interfaces, peripheral nerve stimulation (PNS) of the sensory fibers proximal to limb amputation can restore sensations from the missing extremity in the brain [21–24]. Indeed, a human–machine system based on intraneural electrodes for restoring limb sensations in LLA was successfully developed recently [25–28] (Fig. 1). In this technology, the prosthetic sensors' provide data that are translated into the language of the nervous system of three amputees, achieving significant health, cognitive and functional benefits, as demonstrated during clinical validation. Natural sensory feedback has been restored in transfemoral amputees, which they can use to improve the use of the leg prosthesis during different ambulation tasks and to promote its integration in their body schema. In particular, we designed a neuroprosthetic framework to restore sensory feedback referred on the phantom lower limb of transfemoral amputees and triggered from the bionic leg by stimulating the residual tibial branch of the sciatic nerve through implanted neural interfaces (purposely-designed using advanced computational models [29–31]).

2   Methods

The neuroprosthesis (Fig. 1a) contains a microprocessor-based lower limb prosthesis equipped with sensors under the foot sole and in the knee, a controlling microcomputer, and a stimulating system. The sensors' readouts are acquired and recorded by the wearable insole and then transmitted to microcomputer, which transduces them into instructions for the neural stimulator [32]. The signals from the insole and prosthetic knee sensors are translated in impulses of current, the language of the human nervous system, which are delivered to the residual peripheral nerve through electrodes, implanted transversally into the nerve itself. This is performed in a real-time configuration, with a delay less than 50 ms, which is so brief that users do not perceive it. Then, nature does the rest: the signals from the residual nerves are conveyed to the user's brain, which can perceive what happens at the prosthesis and adjusts walking accordingly [27]. The machine and the body are finally re-connected.

Sensations of touch, pressure, vibration were elicited from more than 20 positions of the phantom foot sole, and of contraction or solicitation from the muscles of the missing leg. Indeed, when the volunteers were blindfolded and asked to recognize touch under the prosthetic foot or the flexion/extension of the prosthetic knee–or the two conditions simultaneously–they achieved an average of more than 80% successful responses.

3   Results

The neural feedback was exploited in active motor tasks, which proved that our approach improved users' mobility (Fig. 2). Thanks to the neuroprosthesis, all of the volunteers could walk over obstacles without the burden of looking at their artificial limb as they walked. Restoring the awareness of the prosthesis allowed the subjects to feel trampled obstacles and to avoid falls. They climbed/descended stairs around 30% faster with sensations restored than without them. Their agility (e.g. tandem walking) was also increased by the restoration of the connection between the brain and the prosthesis [27].

Then, thanks to the full portability and real-time operation of our novel hardware and software system, amputees stepped out from the laboratory to the ecological environment. The subjects' speed also increased during walking over uneven terrain (i.e. sand), which enhanced the device's usability. Our results demonstrate that induced sensory feedback can be integrated at supraspinal levels to restore the missing leg's functional abilities.

Moreover, we found that walking speed and self-reported confidence in the prosthesis increased while mental and physical fatigue decreased for participants during neural sensory feedback compared to the no-stimulation trials. This is an essential result, since the amputees' risk of a heart attack is more than doubled compared to

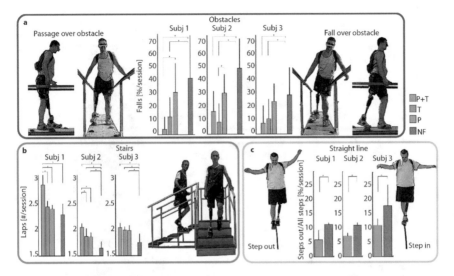

Fig. 2 a Obstacle avoidance task. **b** Stairs Task. **c**. Walking on a straight line. In all these motor tasks, using the neuroprosthetic device restoring artificial sensations to the user guaranteed better mobility relative to using a commercially-available prosthetic leg without sensory feedback. P + T = Proprioceptive + tactile feedback, T = tactile feedback only, P = Proprioceptive feedback only, NF = No Feedback

other people. With continuous use of this system, we assume that this risk will be diminished [25].

Along with the functional and health outcomes, we assessed the cognitive (brain) integration of the device into the body schema of the subjects by measuring the prosthesis embodiment and cognitive effort while using the artificial leg. We also showed increased embodiment of the lower limb prosthesis, through phantom leg displacement perception and questionnaires, and ease of the cognitive effort during a dual-task paradigm (Fig. 3), through electroencephalographic (EEG) recordings. During this task, the subjects had to walk while listening to tones and paying attention to higher ones. Meanwhile, ERP components of the EEG were measured to explore whether walking required more attention with or without sensory feedback. With feedback, users had substantial attention available for other tasks, such as when counting the tones while sitting. However, without feedback, they had limited resources available for other tasks. Therefore, feedback helps amputees walk freely while thinking about activities other than controlling the device. Brain activity measurements and psychophysical tests revealed that the neuroprosthesis is perceived as an extension of the body, like a real limb [27].

Furthermore, participants exhibited reduced phantom limb pain with neural sensory feedback. Indeed, our technology might target both peripheral and central components of pain by using neuromodulation (direct nerve stimulation) and boosting prosthesis cognitive integration (reducing sensorimotor conflicts and distorted phantom limb representations in the brain) [25].

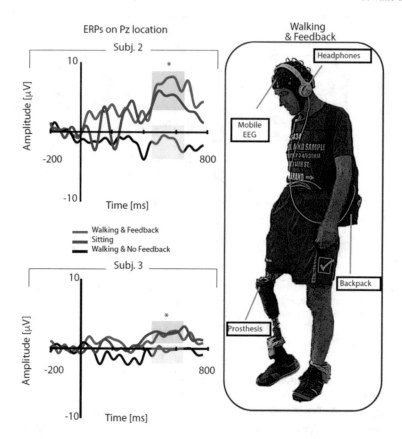

Fig. 3 Dual-task paradigm. The computer-brain interface allowed the prosthetic users to have more mental resources available to perform a cognitive task during walking compared to when the sensory feedback was not provided

Finally, leg amputees are often not satisfied with their prosthesis, even though sophisticated prostheses are becoming available. One important reason for this dissatisfaction is that they perceive the weight of the prosthesis as too high, despite the fact that prosthetic legs are usually less than half the weight of a natural limb. This somatosensory neuroprosthesis connecting the prostheses to the nervous system helped amputees to perceive the prosthesis as lighter, which is beneficial for their acceptance [26].

4   Discussion

The technology behind prosthetic limbs has been advancing rapidly in recent years. Different types of controllable prostheses convert electrical signals from the brain or body into device movements [33–35].

The results from these proof-of-concept cases provide the rationale for larger population studies investigating the clinical utility of neuroprostheses that restore sensory feedback. These works pave the way for further investigations about how the brain interprets different artificial feedback strategies and for the development of fully implantable sensory-enhanced leg neuroprostheses, which could drastically ameliorate quality of life in many people with disabilities.

Notably, we need longer investigations with in-home assessments and a greater number of volunteers to provide more robust data that we can use to draw more significant conclusions. For the time-limited clinical study, signals from the prosthesis were sent along cables through the skin to the electrodes in the thigh. This meant that the volunteers had to undergo regular medical examinations. To eliminate this need, we plan to develop a fully implantable system for the next step. We need to develop a fully-wireless neurostimulation device that can be fully implanted into the patient (like a pacemaker) and bring it to the market [36].

The bionic leg integrated with the residual peripheral nervous system of amputees, namely a computer-brain interface, enables the brain to accept the bionic leg as the continuation of the natural leg. This acceptance is essential for higher confidence of the users, and for future widespread use of these technologies.

Acknowledgements The authors are deeply grateful to the volunteers who freely donated months of their lives for the advancement of knowledge and for a better future for traumatic leg amputees. The authors are also thankful to Prof. Marko Bumbasirevic, Prof. Thomas Stieglitz, Prof. Silvestro Micera, Dr. Asgeir Alexandersson, and Prof. David Guiraud for their support with the surgery and equipment during the experimentation. This project has received funding from the European Research Council (ERC) under the European Union's Horizon 2020 research and innovation program (FeelAgain grant agreement no. 759998), and from H2020-EIC-FTI-2018-2020 GoSafe (grant agreement no. 870144).

References

1. Raspopovic S (2020) Advancing limb neural prostheses. Science 370:290–291. https://doi.org/10.1126/science.abb1073
2. Miller WC, Speechley M, Deathe B (2001) The prevalence and risk factors of falling and fear of falling among lower extremity amputees. Arch Phys Med Rehabil 82:1031–1037. https://doi.org/10.1053/apmr.2001.24295
3. Makin TR, de Vignemont F, Faisal AA (2017) Neurocognitive barriers to the embodiment of technology. Nat Biomed Eng 1s
4. Blanke O (2012) Multisensory brain mechanisms of bodily self-consciousness. Nat Rev Neurosci 13:556–571. https://doi.org/10.1038/nrn3292

5. Walden J (2017) Using administrative healthcare records to identify determinants of amputee residuum outcomes. Walden Dissertations and Doctoral Stud

6. Roffman CE, Buchanan J, Allison GT (2014) Predictors of non-use of prostheses by people with lower limb amputation after discharge from rehabilitation: development and validation of clinical prediction rules. J Physiother 60:224–231. https://doi.org/10.1016/j.jphys.2014.09.003

7. Gailey R (2008) Review of secondary physical conditions associated with lower-limb amputation and long-term prosthesis use. JRRD 45:15–30. https://doi.org/10.1682/JRRD.2006.11.0147

8. Fleury AM, Salih SA, Peel NM (2013) Rehabilitation of the older vascular amputee: a review of the literature. Geriatr Gerontol Int 13:264–273. https://doi.org/10.1111/ggi.12016

9. Hargrove LJ, Simon AM, Young AJ et al (2013) Robotic leg control with EMG decoding in an amputee with nerve transfers. N Engl J Med 369:1237–1242. https://doi.org/10.1056/NEJMoa1300126

10. Hargrove LJ, Young AJ, Simon AM et al (2015) Intuitive control of a powered prosthetic leg during ambulation: a randomized clinical trial. JAMA 313:2244–2252. https://doi.org/10.1001/jama.2015.4527

11. Charkhkar H, Shell CE, Marasco PD et al (2018) High-density peripheral nerve cuffs restore natural sensation to individuals with lower-limb amputations. J Neural Eng 15:056002. https://doi.org/10.1088/1741-2552/aac964

12. Clippinger FW, Seaber AV, McElhaney JH et al (1982) Afferent sensory feedback for lower extremity prosthesis. Clin Orthop Relat Res 202–206

13. Clites TR, Carty MJ, Ullauri JB et al (2018) Proprioception from a neurally controlled lower-extremity prosthesis. Sci Transl Med 10:eaap8373. https://doi.org/10.1126/scitranslmed.aap8373

14. Crea S, Edin BB, Knaepen K et al (2017) Time-discrete vibrotactile feedback contributes to improved gait symmetry in patients with lower limb amputations: case series. Phys Ther 97:198–207. https://doi.org/10.2522/ptj.20150441

15. Rusaw D, Hagberg K, Nolan L, Ramstrand N (2012) Can vibratory feedback be used to improve postural stability in persons with transtibial limb loss? J Rehabil Res Dev 49:1239–1254

16. Dietrich C, Nehrdich S, Seifert S et al (2018) Leg prosthesis with somatosensory feedback reduces phantom limb pain and increases functionality. Front Neurol 9:270. https://doi.org/10.3389/fneur.2018.00270

17. D'Anna E, Valle G, Mazzoni A et al (2019) A closed-loop hand prosthesis with simultaneous intraneural tactile and position feedback. Sci Robot 4:eaau8892. https://doi.org/10.1126/scirobotics.aau8892

18. Nolan L, Wit A, Dudziński K et al (2003) Adjustments in gait symmetry with walking speed in trans-femoral and trans-tibial amputees. Gait Posture 17:142–151. https://doi.org/10.1016/S0966-6362(02)00066-8

19. Waters RL, Perry J, Antonelli D, Hislop HJ (1976) Energy cost of walking of amputees: the influence of level of amputation. J bone joint surg Am 58:42–46. https://doi.org/10.2106/00004623-197658010-00007

20. Srinivasan SS, Tuckute G, Zou J et al (2020) Agonist-antagonist myoneural interface amputation preserves proprioceptive sensorimotor neurophysiology in lower limbs. Sci Transl Med 12:eabc5926. https://doi.org/10.1126/scitranslmed.abc5926

21. Raspopovic S, Capogrosso M, Petrini FM et al (2014) Restoring natural sensory feedback in real-time bidirectional hand prostheses. Sci Transl Med 6:222ra19–222ra19. https://doi.org/10.1126/scitranslmed.3006820

22. Valle G, Mazzoni A, Iberite F et al (2018) Biomimetic intraneural sensory feedback enhances sensation naturalness, tactile sensitivity, and manual dexterity in a bidirectional prosthesis. Neuron 100:37-45.e7. https://doi.org/10.1016/j.neuron.2018.08.033

23. Tan DW, Schiefer MA, Keith MW et al (2014) A neural interface provides long-term stable natural touch perception. Sci Transl Med 6:257ra138. https://doi.org/10.1126/scitranslmed.3008669

24. Ortiz-Catalan M, Hakansson B, Branemark R (2014) An osseointegrated human-machine gateway for long-term sensory feedback and motor control of artificial limbs. Sci Transl Med 6:257re6–257re6. https://doi.org/10.1126/scitranslmed.3008933

25. Petrini FM, Bumbasirevic M, Valle G et al (2019) Sensory feedback restoration in leg amputees improves walking speed, metabolic cost and phantom pain. Nat Med 25:1356–1363. https://doi.org/10.1038/s41591-019-0567-3

26. Preatoni G, Valle G, Petrini FM, Raspopovic S (2021) Lightening the perceived weight of a prosthesis with cognitively integrated neural sensory feedback. Curr Biol 31:1–7. https://doi.org/10.1016/j.cub.2020.11.069

27. Petrini FM, Valle G, Bumbasirevic M et al (2019) Enhancing functional abilities and cognitive integration of the lower limb prosthesis. Sci Transl Med 11:eaav8939. https://doi.org/10.1126/scitranslmed.aav8939

28. Valle G, Saliji A, Fogle E et al (2021) Mechanisms of neuro-robotic prosthesis operation in leg amputees. Sci Adv 7(17). eabd8354

29. Raspopovic S, Petrini FM, Zelechowski M, Valle G (2017) Framework for the development of neuroprostheses: from basic understanding by sciatic and median nerves models to bionic legs and hands. Proc IEEE 105:34–49. https://doi.org/10.1109/JPROC.2016.2600560

30. Zelechowski M, Valle G, Raspopovic S (2020) A computational model to design neural interfaces for lower-limb sensory neuroprostheses. J NeuroEngineering Rehabil 17:24. https://doi.org/10.1186/s12984-020-00657-7

31. Romeni S, Valle G, Mazzoni A, Micera S (2020) Tutorial: a computational framework for the design and optimization of peripheral neural interfaces. Nat Protoc 15:3129–3153. https://doi.org/10.1038/s41596-020-0377-6

32. Valle G, Strauss I, D'Anna E et al (2020) Sensitivity to temporal parameters of intraneural tactile sensory feedback. J NeuroEngineering Rehabil 17:110. https://doi.org/10.1186/s12984-020-00737-8

33. Flesher SN, Collinger JL, Foldes ST et al (2016) Intracortical microstimulation of human somatosensory cortex. Sci Transl Med 8:361ra141–361ra141. https://doi.org/10.1126/scitranslmed.aaf8083

34. Chandrasekaran S, Nanivadekar AC, McKernan G et al (2020) Sensory restoration by epidural stimulation of the lateral spinal cord in upper-limb amputees. eLife 9:e54349. https://doi.org/10.7554/eLife.54349

35. Risso G, Valle G, Iberite F et al (2019) Optimal integration of intraneural somatosensory feedback with visual information: a single-case study. Sci Rep 9:7916. https://doi.org/10.1038/s41598-019-43815-1

36. Raspopovic S, Valle G, Petrini FM (2021) Sensory feedback for limb prostheses in amputees. Nat Mater 1–15

Restoring the Sense of Touch Using a Sensorimotor Demultiplexing Neural Interface: 'Disentangling' Sensorimotor Events During Brain-Computer Interface Control

Patrick D. Ganzer, Samuel C. Colachis 4th, Michael A. Schwemmer, David A. Friedenberg, Collin F. Dunlap, Carly E. Swiftney, Adam F. Jacobowitz, Doug J. Weber, Marcia A. Bockbrader, and Gaurav Sharma

Abstract *Rationale* Paralyzed muscles can be reanimated following spinal cord injury (SCI) using a brain-computer interface (BCI) to enhance motor function alone. Importantly, the sense of touch is a key component of motor function. Simultaneously restoring the sense of touch and movement would meet functional needs for BCI users that seek to enhance upper limb function following SCI. *Methods* The study met institutional requirements for the conduct of human subjects and is registered on the http://www.ClinicalTrials.gov website (identifier: NCT01997125). The participant was a 27-year-old male with stable, non-spastic C5 quadriplegia resulting from a cervical SCI. The participant underwent implantation of a 96 channel Utah microelectrode recording array (Blackrock Microsystems, Inc.; Salt Lake, Utah) in

P. D. Ganzer (✉) · S. C. Colachis 4th · C. F. Dunlap · C. E. Swiftney · A. F. Jacobowitz · D. J. Weber · G. Sharma
Medical Devices and Neuromodulation, Battelle Memorial Institute, 505 King Ave, Columbus, OH 43201, USA
e-mail: pxg487@miami.edu

C. F. Dunlap · M. A. Bockbrader
Center for Neuromodulation, The Ohio State University, Columbus, OH 43210, USA

Department of Physical Medicine and Rehabilitation, The Ohio State University, Columbus, OH 43210, USA

M. A. Schwemmer · D. A. Friedenberg
Advanced Analytics and Health Research, Battelle Memorial Institute, 505 King Ave, Columbus, OH 43201, USA

D. J. Weber
Department of Mechanical Engineering and Neuroscience, Carnegie Mellon University, 5000 Forbes Ave, Pittsburgh, PA 15213, USA

P. D. Ganzer
Department of Biomedical Engineering, University of Miami, Miami, FL 33146, USA

The Miami Project to Cure Paralysis, University of Miami, Miami, FL 33136, USA

© The Author(s), under exclusive license to Springer Nature Switzerland AG 2021
C. Guger et al. (eds.), *Brain-Computer Interface Research*,
SpringerBriefs in Electrical and Computer Engineering,
https://doi.org/10.1007/978-3-030-79287-9_8

his left primary motor cortex. The hand area of motor cortex was identified preoperatively by fusing functional magnetic resonance imaging (fMRI) activation maps obtained while the participant attempted movements co-registered to the preoperative planning MRI. During experiments, the participant was either completely blinded to the experimental conditions or given brief instructions to complete the necessary actions. Cue and trial parameters were randomized as needed. *Results* Results are adapted from (Ganzer PD, Colachis 4th SC, Schwemmer MA, Friedenberg DA, Dunlap CF, Swiftney CE, Sharma G in Restoring the sense of touch using a sensorimotor demultiplexing neural interface. Cell 2020). We demonstrate that a human participant with a clinically complete SCI can use a BCI to simultaneously reanimate both motor function and the sense of touch, leveraging residual touch signaling from their own hand. In primary motor cortex (M1), residual subperceptual hand touch signals are simultaneously demultiplexed from ongoing efferent motor intention, enabling intracortically controlled closed-loop sensory feedback. Using the closed-loop demultiplexing BCI almost fully restored the ability to detect object touch, and significantly improved several sensorimotor functions. *Conclusion* These results demonstrate that subperceptual neural signals can be decoded from human cortex and transformed into conscious perception, significantly augmenting function.

Keywords Spinal cord injury · Upper limb · Touch · Brain-computer interface · Cortex · Machine learning · Decoding · Demultiplex · Sensory feedback

1 Introduction

Spinal cord injury (SCI) damages sensorimotor circuits leading to paralysis, an impaired sense of agency, and sensory dysfunction. Clinical studies are now identifying a new class of SCI—'sensory discomplete'—where tactile stimuli that the patient cannot feel still evoke changes in cortical activity [2–5]. These clinical assessments uncover a set of critical findings. The existence of spared somatosensory fibers, and therefore residual somatosensory information, can potentially be leveraged for functional benefit in patients living with a severe SCI.

We assessed the hypothesis that a BCI could leverage sensory discompleteness, enhance subperceptual touch events, and simultaneously restore both the sense of touch and motor function in a participant with a clinically complete SCI. The study's participant is chronically paralyzed from a clinically complete AIS-A C5 SCI (American Spinal Injury Association Impairment Scale, grade A), and has an intracortical recording array implanted in primary motor cortex (M1) for BCI recordings. During BCI operation, the participant uses his own hand, addressing a major need of patients with SCI [6–8].

Several BCI studies have targeted M1 to decode motor intention alone [9–24, 27]. Motor intention decoded from M1 is then used to enhance motor control via a robotic limb, assistive device, or the participant's own hand via functional electrical stimulation (FES). Sensory discompleteness may allow for touch-related sensory information transmission to the BCI recording site in M1. If so, this residual touch-related sensory information could be used for restoring the sense of touch. Sensory function can potentially be augmented using a BCI that can decipher residual sensory neural activity from the impaired hand and dynamically translate this into closed-loop sensory feedback that the user can perceive.

1.1 Methods

Please refer to [1] for detailed methods and statistical tests related to the study. Below, we provide a brief methods overview.

2 Brief Methods Overview

2.1 Study Participant

Approval for this study was obtained from the US Food and Drug Administration (Investigational Device Exemption) and The Ohio State University Medical Center Institutional Review Board (Columbus, Ohio). The study met institutional requirements for the conduct of human subjects and was registered on the http://www.Clinic alTrials.gov website (identifier: NCT01997125). The participant referenced in this work completed an informed consent process before commencement of the study. The participant was either completely blinded to the experimental conditions or given brief instructions to complete the necessary actions. Cue and trial parameters were randomized as needed, detailed below.

The study participant was a 27-year-old male with stable, non-spastic C5 quadriplegia resulting from a cervical SCI. The participant underwent implantation of a 96 channel Utah microelectrode recording array (Blackrock Microsystems, Inc.; Salt Lake, Utah) in his left primary motor cortex (Fig. 1a). The hand area of motor cortex was identified preoperatively by fusing functional magnetic resonance imaging (fMRI) activation maps obtained while the patient attempted movements co-registered to the preoperative planning MRI. Full details of the fMRI and surgical procedures can be found in Bouton et al. [17]. Neural data were acquired using a Utah microelectrode recording array (Blackrock Microsystems, Inc.; Salt Lake City, Utah) and the Neuroport neural data acquisition system. Recorded data from all 96 recording array channels were sampled at 30 kHz and band pass filtered online from 0.3 to 7.5 kHz using a third order Butterworth analog hardware filter. The neural data was then digitized and sent to a PC for saving or further on-line processing using a custom interface in MATLAB 2014a (The MathWorks; Natick, MA).

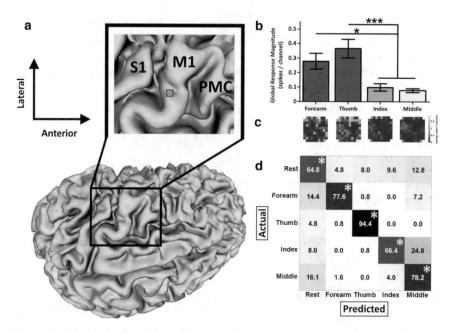

Fig. 1 Skin Stimulation on the Arm and Hand Evokes Robust Neural Responses in Contralateral Primary Motor Cortex (M1) Following Clinically Complete Cervical Spinal Cord Injury (SCI). **a** Reconstruction of the participant's cerebrum and location of BCI implant (red box, inset) in left M1. **b** Magnitude of sensory evoked multiunit activity in M1 following stimulation to 4 skin sites on the right arm and hand. Data presented are mean ± S.E.M. *** = $p < 0.001$, * = $p < 0.05$. **c** Color coded responses across 96 channel Utah array at the 4 different skin stimulation locations. **d** SVM based decoding of stimulation location (or rest) from evoked neural activity. * = above chance at $p < 0.001$

2.2   Brief Methods for Fig. 1

The participant's right hand is largely insensate due to the AIS-A C5 SCI (clinical sensory assessment: Fig. S1, [1]). Passive electrical stimulation occurred on 4 different skin locations (right side: forearm, thumb, index finger, and middle finger) that are either partially intact or completely insensate, while neural activity was recorded from the array implanted in left M1. The participant was blindfolded during the recordings. We used the peristimulus time histogram (PSTH) method to quantify evoked neural activity, similar to previous studies [25, 26]. We report the average response magnitude across all 96 channels of the array during the 4 different recording conditions in Fig. 1b (color coded neural activity across all 96 array channels, Fig. 1c). The location of skin stimulation was decoded using a non-linear support vector machine (SVM; results in Fig. 1d). We used an SVM approach similar to our previous studies [17, 21, 22, 27]. The SVM decoded 5 states/classes [rest, forearm stimulation ('Forearm'), thumb stimulation ('Thumb'), index finger stimulation ('Index'), or middle finger stimulation ('Forearm')]. We present a confusion matrix outlining the SVM's performance.

2.3 Brief Methods for Fig. 2

We created a touch decoder using subperceptual residual neural activity during active object touch (during touch of the 'can object', a part of the standard clinical grasp and release test battery, 5.4 × 9.1 cm; [21, 22, 29]). Active touch cues consisted of a 6 s period. For each touch cue period, the participant first moved his hand down onto and around the can object for 3 s, followed by a scripted object grip period for an additional 3 s, where FES triggered a more forceful grip. A touch decoder SVM was trained and tested on 4 cue types and rest periods to assess model performance during 'touch' and

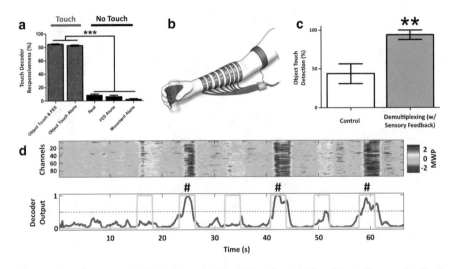

Fig. 2 Active Object Touch Can Be Decoded from M1 to Control Closed-Loop Sensory Feedback and Enhance Hand Sensory Function. **a** Touch decoders were first assessed using 'Touch' or 'No Touch' periods. Touch decoders had significantly higher responsiveness during object touch events (red), compared to control cues lacking object touch (black). Touch decoder false positive rates during cues (data not shown): Object Touch and FES = 12.2%; Object Touch Alone = 13.7%; FES Alone = 3.7%; Movement Alone = 3.3%. **b** Touch decoders next controlled closed-loop sensory feedback via a vibrotactile array interfaced with the sensate skin over the ipsilateral bicep (red band in the cartoon schematic). Closed-loop sensory feedback triggered by residual sensory information in M1 more than doubled object touch detection during object grip (**c**, up to ~ 93%) (** = $p < 0.01$). **d** Representative color-coded mean wavelet power (MWP) input (top) and touch decoder outputs (bottom) during the object touch detection assessment (object placed on cue numbers 2, 4, and 6, # symbol added; cue periods = gray lines; device activation threshold = horizontal dashed line). These results demonstrate that residual subperceptual sensory information can be decoded from M1 to trigger closed-loop tactile feedback and significantly improve sensory function. Data presented are mean ± S.E.M

'no touch' events. The participant separately completed the following cued events:
(1) 3 s of natural touch of the object followed by 3 s of FES mediated touch ('Touch'),
(2) 6 s of natural object touch ('Touch'), (3) 6 s of identical movement without the
object present ('No Touch'), (4) 6 s of FES without the object present ('No Touch').
We report model responsiveness during the 4 cue types and rest (Fig. 2a), defined as
the percentage of time the touch decoder output was above the activation threshold
during the given period. This touch decoder was then used to trigger the closed-loop
sensory feedback interface (interface schematic, Fig. 2b) during the experiments
described in Figs. 2c, d, and 3.

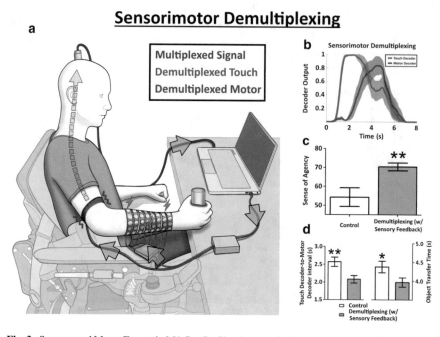

Fig. 3 Sensory and Motor Events in M1 Can Be Simultaneously Decoded to Enable 'Sensorimotor
Demultiplexing' BCI Control and Enhancement of Sensorimotor Function. **a** Schematic of the
participant performing a modified GRT task with the 'sensorimotor demultiplexing' BCI. **b** We
first challenged the touch decoder with a competing simultaneous motor decoder. As expected,
touch decoders were activated before motor decoders on all object transfers (time 0 = touch cue,
followed by participant-initiated motor intention; shaded bands = ± 95% confidence interval of
decoder output). Closed-loop sensory feedback triggered by demultiplexed sensory neural activity
significantly improved the participant's sense of agency (**c**), motor decoder latency (**d**, left), and
object transfer time (**d**, right) (average number of objects transferred per GRT assessment block:
control = 9, demultiplexing with sensory feedback = 9.75). These results demonstrate the ability
to decode afferent and efferent information from M1 and activate multiple assistive devices for
augmenting sensorimotor function, constituting a 'sensorimotor demultiplexing' BCI (* = $p <
0.05$; ** = $p < 0.01$). Data presented are mean ± S.E.M

2.4   Brief Methods for Fig. 3

'Sensorimotor demultiplexing' BCI control was next implemented for multidevice control (i.e., FES and closed-loop haptic feedback) and assessment of upper limb function (Fig. 3a). In a subset of experiments, the participant was cued to first touch the 'can object', and approximately 1 s later think of movement (representative averaged SVM decoder outputs are shown in Fig. 3b, demonstrating simultaneous decoding of residual touch and movement related neural activity). Related to Fig. 3c, the participant was cued to repeatedly grasp, move, and release the can object during shuffled series of 'Demultiplexing With Sensory Feedback' or 'Control' condition trials. After each GRT trial, the participant reported his sense of agency (SoA) (i.e., "How in control did you feel of the movement and grip?"). The SoA score ranged from 0 to 100, similar to previous studies [28, 30] (0 = poor sense of control; 100 = perfect sense of control). Related to Fig. 3d, the participant engaged in a modified GRT using the can object. The participant was instructed to repeatedly grasp, transfer, and release the can object onto an elevated platform as fast as possible during shuffled series of 'Demultiplexing With Sensory Feedback' or 'Control' condition trials. Each GRT assessment period consisted of two 60 s object transfer periods separated by a 20 s rest period. All GRT trials were recorded with high-speed video for offline analysis. We quantified the number of objects successfully transferred and the object transfer times, similar to our previous studies [21, 22]. A successful transfer started the moment the object was initially contacted by the hand and ended when the object was fully released onto the platform (no objects were dropped). We also assessed the interval between the touch decoder and motor decoder activations to examine the neurophysiological substrates of GRT performance with and without sensory feedback (high-speed video was also used in addition to decoder times to confirm touch and motor event start times). The touch decoder or motor decoder start times were calculated across GRT trials using the time each decoder crossed the device activation threshold (device activation threshold = 0.5). We report the interval (s) between the touch and motor decoder activations across testing conditions.

3   Results

The study participant uses his hand during limb reanimation. Unfortunately, his hand is almost completely insensate due to the severe AIS-A C5 SCI (only residual abnormal sensation remaining on the thumb; see [1]: Fig. S1). We first assessed whether any residual sensory information could significantly modulate neural activity at the M1 BCI site (Fig. 1a) following skin stimulation. Sensory stimuli on the arm and hand that the participant can and cannot feel significantly modulated M1 activity (Fig. 1b, c; four skin locations above, at, or below the AIS-A C5 SCI; $F[3,380] = 9.8$, $p < 0.001$). Furthermore, the location of sensory stimulation could be decoded using a non-linear support vector machine (SVM; Fig. 1d). These results demonstrate the ability to both evoke and decode residual sensory neural activity from M1 that is below conscious perception, from functionally relevant hand dermatomes. Overall, this intracortical electrophysiological evidence of sensory discompleteness extends previous studies using noninvasive imaging of evoked activity [2–5].

We next used these residual hand sensory signals in M1 to control real-time closed-loop sensory feedback (Fig. 2). Active touch decoders were constructed from M1 neural activity to detect object touch with high responseiveness (Fig. 2a; F[4,85] = 777, $p < 0.001$; task training cartoon: Fig. 2b, red hand dermatome touching the can object during a standardized clinical assessment). While blindfolded, the participant was unable to decipher object touch above chance (Fig. 2c, white bar). Closed-loop sensory feedback triggered by residual hand touch signals (Fig. 2d, via haptic feedback on sensate skin) almost fully restored the ability to detect object touch (Fig. 2c, gray bar, t(30) = 3.5, $p = 0.001$; representative decoder inputs and outputs: Fig. 2d). These results demonstrate that subperceptual sensory neural activity during active touch can be decoded from M1 and enhanced into conscious perception for functional benefit.

Our final set of experiments assessed the hypothesis that afferent and efferent activity in M1 can be demultiplexed to simultaneously control devices for sensory feedback and FES, constituting a 'sensorimotor demultiplexing' BCI (Fig. 3a). The touch decoder controls closed-loop vibrotactile sensory feedback for enhancing subperceptual hand touch events (red band on bicep, Fig. 3a). The motor decoder simultaneously controls FES of the arm to produce the desired hand movement (blue bands on forearm, Fig. 3a). Real-time 'sensorimotor demultiplexing' was demonstrated during a modified grasp and release test (GRT; [29]).

Residual touch (Fig. 3b, red) and motor intention signals (Fig. 3b, blue) were reliably demultiplexed in real-time (during cued events, while the participant inter-acted with the standarized 'can object'). This result demonstrates that the touch decoder is not significantly impacted by neural activity from simultaneous move-ment intention events. 'Sensorimotor demultiplexing' BCI control was next enabled using the simultaneous decoding of touch and motor intention events during a set of upper limb assessments. This closed-loop 'sensorimotor demultiplexing' BCI system enabled significant improvements in sense of agency (Fig. 3c), BCI system speed (Fig. 3d, left), and object transfer time (Fig. 3d, right) compared to a motor-only BCI control. Therefore, rapid closed-loop sensory feedback not only augments sensory function, but also augments motor function. These findings demonstrate a BCI system that simultaneously demultiplexes afferent and efferent activity from cortex for controlling multiple assistive devices and enhancing function.

4   Brief Discussion

Severe AIS-A SCI should essentially eliminate sensory information transmission to the brain that originates from skin innervated from below the lesion. Recent studies demonstrate that residual subperceptual sensory information from below the lesion is transmitted to sensory areas of the brain, even following severe clinically complete SCI ('sensory discompleteness': [2–5]). We extend these results, and show that sensory discompleteness can be leveraged by a BCI for improvement in function. The sensorimotor demultiplexing capability can impact how BCI electrode array implant locations are determined for interfaces seeking to decode multiplexed information classes relevant for BCI control. For future BCIs, it will be critical to perform multimodal pre-surgical brain mapping to localize these relevant neural representations and further inform electrode array implant location. Overall, our results support the hypothesis that subperceptual residual neural information can be reliably decoded from the human brain, and used to augment function.

Funding Financial support for this study came from Battelle Memorial Institute and The Ohio State University Center for Neuromodulation.

References

1. Ganzer PD, Colachis SC IV, Schwemmer MA., Friedenberg DA, Dunlap CF, Swiftney CE, Sharma G (2020) Restoring the sense of touch using a sensorimotor demultiplexing neural interface. Cell
2. Wrigley PJ, Siddall PJ, Gustin SM (2018) New evidence for preserved somatosensory pathways in complete spinal cord injury: a fMRI study. Hum Brain Mapp 39(1):588–598
3. Awad A, Levi R, Lindgren L, Hultling C, Westling G, Nyberg L, Eriksson J (2015) Preserved somatosensory conduction in a patient with complete cervical spinal cord injury. J Rehabil Med 47(5):426–431. https://doi.org/10.2340/16501977-1955
4. Ioannides AA, Liu L, Khurshudyan A, Bodley R, Poghosyan V, Shibata T, Dammers J, Jamous A (2002) Brain activation sequences following electrical limb stimulation of normal and paraplegic subjects. Neuroimage 16(1):115–129
5. Sabbah P, De SS, Leveque C, Gay S, Pfefer F, Nioche C, Sarrazin JL, Barouti H, Tadie M, Cordoliani YS (2002) Sensorimotor cortical activity in patients with complete spinal cord injury: a functional magnetic resonance imaging study. J Neurotrauma 19(1):53–60
6. Anderson KD (2004) Targeting recovery: priorities of the spinal cord-injured population. J Neurotrauma 21:1371–1383
7. Snoek GJ, IJzerman MJ, Hermens HJ, Maxwell D and Biering-Sorensen F (2004) Survey of the needs of patients with spinal cord injury: impact and priority for improvement in hand function in tetraplegics. Nat Spinal Cord 42 526–32
8. Blabe CH, Gilja V, Chestek CA, Shenoy KV, Anderson KD, Henderson JM (2015) Assessment of brain–machine interfaces from the perspective of people with paralysis. J Neural Eng 12:1–9
9. Lebedev MA, Nicolelis MAL (2017) Brain-machine interfaces: from basic science to neuroprostheses and neurorehabilitation. Physiol Rev 97:767–837
10. Hochberg LR et al (2012) Reach and grasp by people with tetraplegia using a neurally controlled robotic arm. Nature 485:372–375
11. Collinger JL et al (2013) High-performance neuroprosthetic control by an individual with tetraplegia. The Lancet 381:557–564
12. Gilja V et al (2015) Clinical translation of a high-performance neural prosthesis. Nat Med 21:1142–1145
13. Simeral JD, Kim SP, Black MJ, Donoghue JP, Hochberg LR (2011) Neural control of cursor trajectory and click by a human with tetraplegia 1000 days after implant of an intracortical microelectrode array. J Neural Eng 8:025027

14. Jarosiewicz B et al (2015) Virtual typing by people with tetraplegia using a self-calibrating intracortical brain–computer interface. Sci Transl Med 7:313ra179
15. Bockbrader MA, Francisco G, Lee R, Olson J, Solinsky R, Boninger ML (2018) Brain computer interfaces in rehabilitation medicine. PM&R. 10(9S2):S233-S243. doi: https://doi.org/10.1016/j.pmrj.2018.05.028
16. Moxon KA, Foffani G (2015) Brain-machine interfaces beyond neuroprosthetics. Neuron 86(1):55–67. https://doi.org/10.1016/j.neuron.2015.03.036
17. Bouton CE, Shaikhouni A, Annetta NV, Bockbrader MA, Friedenberg DA, Nielson DM, Sharma G, Sederberg PB, Glenn BC, Mysiw WJ, Morgan AG, Deogaonkar M, Rezai AR (2016) Restoring cortical control of functional movement in a human with quadriplegia. Nature 533(7602):247–250
18. Friedenberg DA, Schwemmer MA, Landgraf AJ, Annetta NV, Bockbrader MA, Bouton CE, Zhang M, Rezai AR, Mysiw WJ, Bresler HS, Sharma G (2017) Neuroprosthetic enabled control of graded arm muscle contraction in a paralyzed human. Sci Rep 7(1):8386
19. Sharma G, Friedenberg DA, Annetta N, Glenn B, Bockbrader M, Majstorovic C, Domas S, Mysiw WJ, Rezai A, Bouton C (2016) Using an artificial neural bypass to restore cortical control of rhythmic movements in a human with quadriplegia. Sci Rep 6:33807. https://doi.org/10.1038/srep33807
20. Skomrock ND, Schwemmer MA, Ting JE, Trivedi HR, Sharma G, Bockbrader MA, Friedenberg DA (2018) A characterization of brain-computer interface performance trade-offs using support vector machines and deep neural networks to decode movement intent. Front Neurosci 12:763. https://doi.org/10.3389/fnins.2018.00763
21. Schwemmer MA, Skomrock ND, Sederberg PB, Ting JE, Sharma G, Bockbrader MA, Friedenberg DA (2018) Meeting brain-computer interface user performance expectations using a deep neural network decoding framework. Nat Med 24:1669–1676. https://doi.org/10.1038/s41591-018-0171-y
22. Colachis SC 4th, Bockbrader MA, Zhang M, Friedenberg DA, Annetta NV, Schwemmer MA, Skomrock ND, Mysiw WJ, Rezai AR, Bresler HS, Sharma G (2018) Dexterous control of seven functional hand movements using cortically-controlled transcutaneous muscle stimulation in a person with tetraplegia. Front Neurosci 12:208. https://doi.org/10.3389/fnins.2018.00208
23. Ajiboye AB et al (2017) Restoration of reaching and grasping movements through brain-controlled muscle stimulation in a person with tetraplegia: a proof-of-concept demonstration. The Lancet. https://doi.org/10.1016/S0140-6736(17)30601-3
24. Bockbrader M, Annetta N, Friedenberg D, Schwemmer M, Skomrock N, Colachis S 4th, Zhang M, Bouton C, Rezai A, Sharma G, Mysiw WJ (2019) Clinically significant gains in skillful grasp coordination by an individual with tetraplegia using an implanted brain-computer interface with forearm transcutaneous muscle stimulation. Arch Phys Med Rehabil. pii:S0003–9993(19)30163–7. doi: https://doi.org/10.1016/j.apmr.2018.07.445
25. Manohar A, Foffani G, Ganzer PD, Bethea J, Moxon KA (2017) Cortex dependent recovery of unassisted hindlimb locomotion after complete spinal cord injury in the adult rat. eLife 6 pii: e23532
26. Ganzer PD, Moxon KA, Knudsen EB, Shumsky JS (2013) Serotonergic pharmacotherapy promotes cortical reorganization after spinal cord injury. Exp Neurol 241:84–94
27. Annetta N, Friend J, Schimmoeller A, Buck VS, Friedenberg D, Bouton CE, Bockbrader MA, Ganzer P, Colachis S, Zhang M, Mysiw WJ, Rezai AR, Sharma G (2018) A high definition non-invasive neuromuscular electrical stimulation system for cortical control of combinatorial rotary hand movements in a human with tetraplegia. IEEE Trans Biomed Eng. https://doi.org/10.1109/TBME.2018.2864104
28. Obhi SS, Hall P (2011) Sense of agency and intentional binding in joint action. Exp Brain Res 211:655–662

29. Wuolle KS, Doren CLV, Thrope GB, Keith MW, Peckham PH (1994) Development of a quantitative hand grasp and release test for patients with tetraplegia using a hand neuroprosthesis. J Hand Surg 19:209–342
30. Kumar N, Manjaly JA, Miyapuram KP (2014) Feedback about action performed can alter the sense of self-agency. Front Psychol 5:145

A Brain-Spine Interface Complements Deep-Brain Stimulation

Tomislav Milekovic

Abstract This chapter presents an interview with Tomislav Milekovic, who led a large multinational team that developed a brain-spine interface to help patients with Parkinson's disease (PD). This team submitted their work to the BCI Research Awards in 2020 and won second place. Their interface worked in synergy with deep brain stimulation (DBS) and could potentially help patients who have difficulty with gait and balance, which are major problems in PD. This interview presents results from a primate model, and their group is now working toward clinical trials in humans.

1 Introduction

More than 90% of people with Parkinson's disease (PD) suffer from gait and balance deficits that reduce quality of life. These deficits are associated with disrupted communication between the brain and spinal cord resulting from the depletion of dopaminergic and cholinergic circuits in PD. In late-stage PD, commonly available therapies such as sensory cuing, dopamine replacement strategies and deep brain stimulation (DBS) typically cannot overcome these and other deficits. A brain-spine interface could complement other therapies to help patients with PD.

T. Milekovic (✉)
School of Life Sciences, Center for Neuroprosthetics (CNP) and Brain Mind Institute, Swiss Federal Institute of Technology (EPFL), Lausanne, Switzerland
e-mail: tomislav.milekovic@epfl.ch

Department of Clinical Neuroscience, Lausanne University Hospital (CHUV) and, University of Lausanne (UNIL), Lausanne, Switzerland

Defitech Center for Interventional Neurotherapies (NeuroRestore), CHUV/UNIL/EPFL, Lausanne, Switzerland

Department of Fundamental Neuroscience, Faculty of Medicine, University of Geneva, Geneva, Switzerland

C. Guger et al. (eds.), *Brain-Computer Interface Research*,
SpringerBriefs in Electrical and Computer Engineering,
https://doi.org/10.1007/978-3-030-79287-9_9

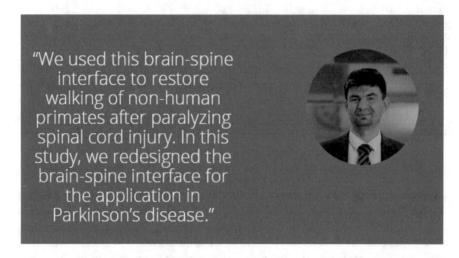

"We used this brain-spine interface to restore walking of non-human primates after paralyzing spinal cord injury. In this study, we redesigned the brain-spine interface for the application in Parkinson's disease."

Their project entailed over two dozen authors from sixteen institutes in the US, China, and different areas of the EU. This was the largest project among all nominees for the 2020 BCI Research Awards in terms of the number of authors and affiliations. The full title, authors, and affiliations were:

A Brain–Spine Interface Complements Deep-Brain Stimulation to both Alleviate Gait and Balance deficits and Increase Alertness in a Primate Model of Parkinson's Disease

Tomislav Milekovic[1,2,3,4], Flavio Raschellà[1,2,3,5], Matthew G. Perich[2], Eduardo Martin Moraud[1,2,3,6], Shiqi Sun[1,2,3,7], Giuseppe Schiavone[8], Yang Jianzhong[9,10], Andrea Galvez[1,2,3,4], Christopher Hitz[1], Alessio Salomon[1], Jimmy Ravier[1,2,3], David Borton[1,11], Jean Laurens[1,12], Isabelle Vollenweider[1], Simon Borgognon[1,2,3], Jean-Baptiste Mignardot[1], Wai Kin D Ko[9,10], Cheng YunLong[9,10], Li Hao[9,10], Peng Hao[9,10], Laurent Petit[13,14], Qin Li[9,10], Marco Capogrosso[1], Tim Denison[15], Stéphanie P. Lacour[8], Silvestro Micera[5,16], Chuan Qin[10], Jocelyne Bloch[1,2,3,6], Erwan Bezard[9,10,13,14], Grégoire Courtine[1,2,3,6]

[1] *Center for Neuroprosthetics (CNP) and Brain Mind Institute, School of Life Sciences, Swiss Federal Institute of Technology (EPFL), Switzerland*

[2] *Department of Clinical Neuroscience, Lausanne University Hospital (CHUV) and University of Lausanne (UNIL), Switzerland*

[3] *Defitech Center for Interventional Neurotherapies (NeuroRestore), CHUV/UNIL/EPFL, Switzerland*

[4] *Department of Fundamental Neuroscience, Faculty of Medicine, University of Geneva, Switzerland*

[5] *CNP and Institute of Bioengineering, School of Engineering, EPFL, Switzerland*

[6] *Department of Neurosurgery, CHUV, Switzerland*

[7] *Beijing Engineering Research Center for Intelligent Rehabilitation, College of Engineering, Peking University, People's Republic of China*

[8] *CNP, Institute of Microengineering and Institute of Bioengineering, School of Engineering, EPFL, Switzerland*

[9] *Motac Neuroscience, UK*

[10] *Institute of Laboratory Animal Sciences, China Academy of Medical Sciences, People's Republic of China*

[11] *Carney Institute for Brain Science, School of Engineering, Brown University, USA*

[12] *Department of Neuroscience, Baylor College of Medicine, USA*

[13] *Université de Bordeaux, Institut des Maladies Neurodégénératives (IMN), UMR 5293, France*

[14] *CNRS, IMN, UMR 5293, France*

[15] *Oxford University, UK*

[16] *The BioRobotics Institute, Scuola Superiore Sant'Anna, Italy*

This project won second place in the 2020 BCI Research Awards in our first tie for second place. This book includes a chapter with an interview with the first author of the other second-place winning project, Dr. Moly. That project also involved a BCI system to help people who have difficulty moving, although their system was different in many other ways.

2 Interview

Hi Tomislav, you submitted your BCI research "A Brain-Spine Interface Complements Deep-Brain Stimulation to Both Alleviate Gait and Balance Deficits and Increase Alertness in a Primate Model of Parkinson's Disease" to the BCI Award 2020 and won 2nd place. Could you briefly describe what this project was about?

Tomislav: We previously developed a brain–spine interface—a neuroprosthesis that infers movement intentions from cortical activity and then modulates spinal cord stimulation to elicit or reinforce those movements. We used this brain-spine interface to restore walking of non-human primates after paralyzing spinal cord injury. Somewhat similar to spinal cord injury, Parkinson's disease also disrupts the communication between the brain and spinal cord. As a result, the Parkinson's disease patients can experience pronounced gait and balance deficits. A cure or effective palliative treatment for these patients is still missing. In this study, we redesigned the brain-spine interface for the application in Parkinson's disease. We demonstrated that this neuroprosthesis substantially alleviated Parkinsonian gait and balance deficits in a non-human primate model of the disease. Furthermore, we showed that the

Fig. 1 Prof. Gregoire Courtine, who co-directs the NeuroRestore center together with Prof. Joce-lyne Bloch and is the lead senior author of the study, displays the main components of the brain-spine interface—a neuroprosthetic system that directly links the brain with the spinal cord to alle-viate neurological motor deficits. All devices needed for the clinical brain-spine interface are FDA approved for medical use, or have an FDA IDE approval and are currently being used in human clinical trials. Preparations for a clinical trial to demonstrate safety and efficacy of the brain-spine interface in people with Parkinson's are underway

brain-spine interface synergizes with the deep brain stimulation, which is a standard clinical therapy for late-stage Parkinson's disease patients (Fig. 1).

What was your goal?
Tomislav: Our goal was to develop and demonstrate a therapy that can alleviate gait and balance deficits of Parkinson's disease patients, a condition that currently has no effective treatments (Fig. 2).

What technologies did you use?
Tomislav: We assembled a brain-computer interface system that comprised:

1. Blackrock microelectrode arrays implanted into the left and right leg motor cortex to record action potentials of motor cortical neurons.
2. A skull mounted pedestal and a wireless data transmission module used to send the neural signals to an external receiver to allow unimpeded behavior of the subjects.
3. Blackrock Neural Signal Processor to acquire the wirelessly received neural signals.

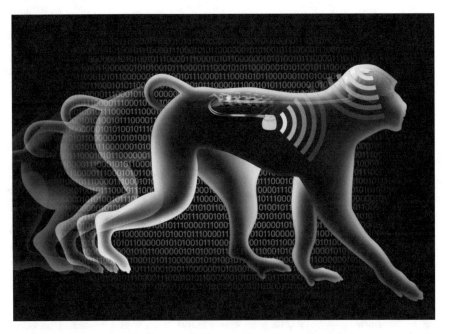

Fig. 2 Illustration of the brain-spine interface validated in a non-human primate model of Parkinson's disease. This neuroprosthesis infers movement attempts from the brain activity wirelessly acquired by a neurosensor. The neuroprosthesis then send wireless stimulation commands to the spinal implant. The spinal implant stimulates the spinal cord to reinforce attempted movements and, therefore, alleviate gait and balance deficits of Parkinson's disease

4. A control computer that ran a software application developed in our lab. This application acquired the neural signals, processed them, inferred the intended movements and then sent out commands to the spinal cord stimulation system to stimulate the spinal cord in a way that reinforces intended movements.
5. A Medtronic software application that ran on the control computer that wirelessly relayed the stimulation commands towards the subject.
6. A Medtronic Patient Programmer that received the wireless stimulation commands and passed them on to through the skin.
7. A Medtronic Activa RC implanted pulse generator that received the stimulation commands and elicited the stimulation.
8. Two in-lab designed and fabricated 8-electrode spinal leads implanted over the dura of the lumbosacral spinal cord that delivered the stimulation.

Apart from the spinal lead, which had to be designed specifically for non-human primates in order to fit within the much smaller spinal canal when compared to humans, all other devices are FDA approved for medical use, or have an FDA IDE approval and are currently being used in human clinical trials. We intentionally relied on medical grade devices to provide a rapid path towards clinical use of our system (Fig. 3).

Fig. 3 The critical components of the brain-spine interface. Right hand: a 96-channel microelectrode Blackrock array is implanted into the phantom of the brain in the area corresponding to the leg motor cortex. These arrays are used to record action potentials of cortical neurons that represent movement attempts. The brain-spine interface acquires these brain signals to infer leg movement attempts. Left hand: a Medtronic implantable pulse generator connects to a multielectrode lead to form a spinal implant. Brain-spine interface uses inferred movement attempts to wirelessly control the spinal implant in real time. The spinal implant stimulates the spinal cord to reinforce the attempted movements, and therefore alleviate gait and balance deficits of Parkinson's disease

What kinds of people could benefit from your research?
Tomislav: People with Parkinson's disease that develop gait and balance deficits. These deficits are common at the late stage of the disease and are one of the lead causes of injury for Parkinson's disease patients. Both the deficits and the injuries that result from them severely impact the quality of life for these patients.

Do you think your work as future potential for clinical use?
Tomislav: Our brain-spine interface relied on medical grade devices. This gives us a clear path towards clinical trials and subsequent use of the technology around the world. We are currently preparing applications for regulatory approval to conduct the first-in-human clinical trials with a therapy system that was derived from the brain-spine interface described in our project. We are looking forward to demonstrating the efficacy of this system in patients with Parkinson's disease in the near future.

What was it like to win the second place in the BCI Award 2020?
Tomislav: Exhilarating. I've spent my whole research career (10+ years) in the BCI field and was elated to receive this recognition of my work from the BCI research

community. Equally important, it was great to meet with the BCI community, especially since this year's BCI social at the SFN conference could not be held. I do hope to see everyone again in the coming years.

How can students and other researchers get involved in your research?
Tomislav: My host institution, the Defitech Center for Interventional Neurotherapies—NeuroRestore—is engaged in several cutting-edge BCI clinical trials. The center is nested in Lausanne, Switzerland, which features beautiful views over the Lake Geneva and has amongst the highest academic salaries worldwide. We have several open positions and are very much open for collaborations. They can find more details at www.neurorestore.swiss.

Automatic Speech Separation Enables Brain-Controlled Hearable Technologies

Cong Han, James O'Sullivan, Yi Luo, Jose Herrero, Ashesh D. Mehta, and Nima Mesgarani

Abstract People with hearing impairment have difficulty hearing a speaker's voice amidst competing sound sources. While traditional hearing aids can suppress background noise, they cannot help a user listen to a single conversation among many without knowing which speaker the user is attending to. In this work, we design a brain-controlled hearing aid that can automatically determine the speaker that the user is focusing on and amplify that speaker. We propose a novel speech separation algorithm to automatically separate speakers in mixed audio without any need for prior training on the speakers. The separated speakers are compared to evoked neural responses in the auditory cortex of the listener to determine and amplify the attended speaker. We demonstrate that the proposed brain-controlled hearing aid significantly improves speech perception of the attended speaker. By combining the latest advances in speech processing technologies and brain-computer interfaces, the brain-controlled hearing aid can assist individuals with hearing impairments and reduce the listening effort for normal hearing subjects in adverse acoustic environments.

Keywords Speech separation · Auditory attention decoding · Deep learning · Hearing aid · Brain-computer interface (BCI)

Cong Han and James O'Sullivan contributed equally to this work

C. Han · J. O'Sullivan · Y. Luo · N. Mesgarani (✉)
Department of Electrical Engineering, Columbia University, New York, NY, USA
e-mail: nima@ee.columbia.edu

Zuckerman Mind Brain Behavior Institute, Columbia University, New York, NY, USA

J. Herrero · A. D. Mehta
Department of Neurosurgery, Hofstra-Northwell School of Medicine and Feinstein Institute for Medical Research, Manhasset, New York, NY 11030, USA

1 Introduction

Speech communication in acoustic environments with more than one speaker can be extremely challenging for hearing impaired listeners [1]. Assistive hearing devices have seen significant progress in suppressing background noises that are acoustically different from speech [2, 3], but they cannot enhance a target speaker without knowing which speaker the listener is conversing with [4]. Recent discoveries of the properties of speech representation in the human auditory cortex have shown an enhanced representation of the attended speaker relative to unattended sources [5]. These findings have motivated the prospect of a brain-controlled assistive hearing device to constantly monitor the brainwaves of a listener and compare them with sound sources in the environment to determine the most likely talker that a subject is attending to [6]. Then, this device can amplify the attended speaker relative to others to facilitate hearing that speaker in a crowd. This process is termed auditory attention decoding (AAD), a research area that has seen considerable growth in recent years.

Because the attentional focus of the subject is determined by comparing the brainwaves of the listener with each sound source, a practical AAD system needs to automatically separate the sound sources in the environment to detect the attended source and subsequently amplify it. One of the difficulties is speaker-independent speech separation [7], meaning the processing must be generalized to new, unseen speakers when the subject converses with a new person. In recent years, several deep neural network-based methods have been proposed to address this problem [8, 9, 10]. However, they were proposed for non-causal speech separation which required an entire utterance to perform the separation, which limited real-time applications, such as in a hearing device. To alleviate this limitation, we propose a causal, speaker-independent automatic speech separation algorithm, online deep attractor network (ODAN), which can separate unseen speakers with low latency. By combining ODAN and AAD, we introduce a speaker-independent AAD system without clean sources, as shown in Fig. 1. Because this system can generalize to new speakers, it overcomes a major limitation of the previous AAD approach that required training on the target speakers [9]. The proposed AAD framework enhances the subjective and objective quality of perceiving the attended speaker in a multi-talker mixture.

By combining recent advances in automatic speech processing and brain-computer interfaces, this study represents a major advancement toward solving one of the most difficult barriers in actualizing AAD. This solution can help people with hearing impairment communicate more easily.

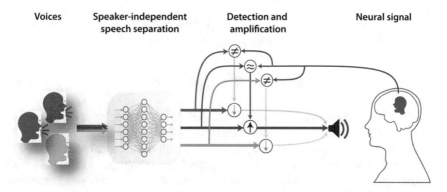

Fig. 1 Schematic of the proposed brain-controlled assistive hearing device. A brain-controlled assistive hearing device can automatically amplify one speaker among many. A deep neural network automatically separates each of the speakers from the mixture and compares each speaker with the neural data from the user's brain to accomplish this goal. Then, the speaker that best matches the neural data is amplified to assist the user

2 Methods

2.1 Speaker-Independent Speech Separation

The problem of speech separation is formulated as estimating C sources, $s_1(t), \ldots, s_c(t) \in \mathbf{R}^{1 \times T}$ from the mixture waveform $x(t) \in \mathbf{R}^{1 \times T}$:

$$x(t) = \sum_{i=1}^{C} s_i(t) \tag{1}$$

Taking the short-time Fourier transform (STFT) of both sides formulates the source separation problem in the time–frequency (T-F) domain, where the complex mixture spectrogram is the sum of the complex source spectrograms:

$$X(f, t) = \sum_{i=1}^{C} S_i(f, t) \tag{2}$$

where $X(f, t)$ and $S_i(f, t) \in C^{F \times T}$. One common approach for recovering the individual sources, S_i, is to estimate a real-valued time–frequency mask for each source, $M_i \in \mathbf{R}^{F \times T}$ such that

$$\left| \hat{S}_i(f, t) \right| = |X(f, t)| \, M_i(f, t) \tag{3}$$

The waveforms of the separated sources are then approximated using the inverse STFT of $\left|\hat{S}_i(f, t)\right|$ using the phase of the mixture audio:

$$\hat{s}_i(t) = IFFT\left(\left|\hat{S}_i(f, t)\right| \angle X(f, t)\right) \tag{4}$$

We design online deep attractor network (ODAN) to estimate the mask for each source from the mixture. In ODAN framework, source separation is performed by first projecting the mixture spectrogram onto a high-dimensional space where T-F bins belonging to the same source are placed closer together to facilitate their assignment to the corresponding sources. This procedure is performed in multiple steps. First, the mixture magnitude spectrogram, $|X(f, t)|.$, is projected onto a tensor, $V(f, t, k)$, where each time–frequency bin is represented by a vector of length K. We refer to this representation as the embedding space. The neural network that embeds the spectrogram consists of a four-layer long short-term memory (LSTM) network followed by a fully connected layer. To assign each embedded T-F bin to one of the speakers in the mixture, we track the centroid of the sakers in the embedding space along time. We refer to the centroids of the source i and at time step τ as the attractor points, $A_{\tau,i}(k)$, because they pull together and attract all the embedded T-F bins that belong to the same source. Therefore, the distance (defined as the dot product [11]) between the embedded T-F bins to each of the attractor points determines the source assignment for that T-F bin, which is then used to construct a mask to recover that source.

$$M_{\tau,i}(f) = Softmax\left(\sum_k A_{\tau,i}(k)\, V_\tau(f, k)\right) \tag{5}$$

where the *Softmax* function is defined as:

$$Softmax(x_i) = \frac{e^{x_i}}{\sum_{i=1}^{C} e^{x_i}}$$

The masks subsequently multiply by the mixture magnitude spectrogram to estimate the magnitude spectrograms of each source (Eq. 3). Figure 2 shows the flowchart of the complete operation of the ODAN system.

2.2 Behavioral AAD Experiment and Neural Measurements Neural Recordings

To test the feasibility of using the ODAN speech separation network in a cognitively controlled hearing device, we used invasive electrophysiology to measure neural activity from three neurosurgical patients undergoing treatment for epilepsy. Two

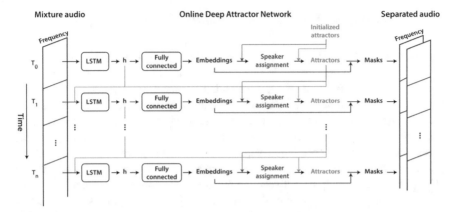

Fig. 2 The architecture of ODANet. In the first frame, attractors are estimated with the help of anchor points. In subsequent frames, the attractor points are updated by dynamically merging the previous and current attractors estimates

subjects (subjects 1 and 2) were implanted with high-density subdural electrocorticography (ECoG) arrays over their language dominant temporal lobe, providing coverage of the superior temporal gyrus (STG), which selectively represents attended speech [5]. The third subject was implanted with bilateral stereotactic EEG (sEEG) with depth electrodes in Heschl's gyrus (HG; containing primary auditory cortex) and STG. This implantation resulted in varying amounts of coverage over the left and right auditory cortices of each subject. All subjects had self-reported normal hearing and consented to participate in the experiment.

Each subject participated in the following experiments for this study: single-talker (S-T) and multi-talker (M-T) experiments. In the S-T experiment, each subject listened to four continuous speech stories (each story was 3 min long) for a total of 12 min of speech material. The stories were uttered once by a female and once by a male speaker (hereafter referred to as Spk1 and Spk2, respectively). For the M-T experiment, subjects were presented with a mixture of the same speech stories as those in the S-T experiment where both speakers were combined at a 0 dB target-to-masker ratio. The M-T experiment was divided into 4 behavioral blocks, each containing a mixture of 2 different stories spoken by Spk1 and Spk2. Before each experimental block, subjects were instructed to focus their attention on one speaker and to ignore the other. All subjects began the experiment by attending to the male speaker and switched their attention to the alternate speaker on each subsequent block. To ensure that subjects were engaged in the task, we intermittently paused the stories and asked subjects to repeat the last sentence of the attended speaker before the pause. All subjects performed the task with high behavioral accuracy and were able to report the sentence before the pause with an average accuracy of 90.5% (S1: 94%, S2: 87%, S3: 90%). Speech sounds were presented using a single loudspeaker placed in front of the subject at a comfortable hearing level, with no spatial separation between the competing speakers.

2.3   Decoding the Listener's Attentional Focus

The reconstructed spectrogram from the auditory cortical responses of a listener in a multi-talker speech perception task is more similar to the spectrogram of the attended speaker than that of the unattended speaker [5]. Therefore, we used a simple classification scheme in which we computed the correlation between the reconstructed spectrograms with both clean attended and unattended speaker spectrograms over a specified duration. Next, the attended speaker is determined as the speaker with a higher correlation value. We used a linear reconstruction method [12] to convert neural responses back to the spectrogram of the sound. This method calculates a linear mapping between the response of a population of neurons to the time–frequency representation of the stimulus [12]. This mapping is performed by assigning a spatiotemporal filter to the set of electrodes, which is estimated by minimizing the MSE between the original and reconstructed spectrograms. We estimated the reconstruction filters using only the neural responses to speech in the S-T experiment. Then, we fixed the filters and used them to reconstruct the spectrogram in the M-T experiments under different attention focuses.

2.4   Psychoacoustic Experiment

To test if the difficulty of attending to the target speaker is reduced using the ODAN-AAD system, we performed a psychoacoustic experiment comparing the original mixture and sounds in which the decoded target speaker was amplified by 12 dB. This particular amplification level has been shown to significantly increase the intelligibility of the attended speaker, while keeping the unattended speakers audible enough to enable attention switching [13]. Subjects were asked to rate the difficulty of attending to the target speaker in three conditions when listening to the following: (1) the raw mixture, (2) enhanced target speech using the output of ODAN-AAD, and (3) enhanced target speech using the output of the Clean-AAD system. Twenty listeners with normal hearing participated in the psychoacoustic experiment where they each heard 20 sentences in each of the three experimental conditions in random order. Subjects were instructed to attend to one of the speakers and report the difficulty of focusing on that speaker. Subjects were asked to rate the difficulty on a scale from 1 to 5 using the mean opinion score (MOS [14]).

3 Results

3.1 Accuracy of Attention Decoding

The duration of the signal used for the calculation of the correlation is an important parameter and affects both the decoding accuracy and speed. Longer durations increase the reliability of the correlation values, hence improving the decoding accuracy. We examined the decoding accuracy with varying duration of the temporal window. The accuracy was calculated for the following cases: when using ODAN spectrograms and when using the actual clean spectrograms. We found no significant difference in decoding accuracy with ODAN or the clean spectrograms when different time windows were used (Wilcoxon rank sum test, P = 0.9). This finding confirms that sources that the ODAN algorithm automatically separates result in the same attention decoding accuracy as obtained with the actual clean spectrograms. As expected, increasing the correlation window resulted in improved decoding accuracy for both ODAN and actual clean sources (Fig. 3a).

Next, we simulated a dynamic switching of attention where the neural responses were concatenated from different attention experiment blocks such that the neural data alternated between attending to the two speakers. We compared the correlation

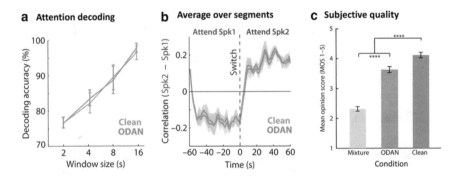

Fig. 3 Evaluating the accuracy of speech separation and attention decoding methods. **a** Attention decoding: The percentage of segments in which the attended speaker was correctly identified for a varying number of correlation window lengths when using ODAN and the actual clean spectrograms. There was no significant difference between using the clean and the ODAN spectrograms (Wilcoxon rank sum test, P = 0.9). **b** Dynamic switching of attention was simulated by segmenting and concatenating the neural data into alternating 60-s bins. The dashed line indicates switching attention. The average correlation values from one subject are shown using a 4-s window size for both ODAN and the actual clean spectrograms. The shaded regions denote SE. **c** Subjective listening test to determine the ease of attending to the target speaker. Twenty healthy subjects were asked to rate the difficulty of attending to the target speaker when listening to (i) the raw mixture, (ii) the ODAN-AAD amplified target speaker, and (iii) the clean-AAD amplified target speaker. The detected target speakers in (ii) and (iii) were amplified by 12 dB relative to the interfering speakers. The bar plots show the median MOS ± SE for each condition. The enhancement of the target speaker for the ODAN-AAD and clean-AAD systems was 100 and 118%, respectively (P < 0.001)

values between the reconstructed spectrograms with both ODAN and the actual clean spectrograms using a sliding window of 4 s. Then, we averaged the correlation values over the segments by aligning them according to the time of the attention switch. Figure 3b shows the average correlation for one example subject over all the segments where the subject was attending to Spk1 in the first 60 s and switched to Spk2 afterward. The overlap between the correlation plots calculated from ODAN and the actual clean spectrograms shows that the temporal properties of attention decoding are the same in both cases; hence, ODAN outputs can replace the clean spectrograms without any significant decrease in decoding speed.

3.2   Increased Subjective Quality of the Attended Speaker

The bar plots in Fig. 3c show the median MOS $\pm$ standard error (SE) for each of the three conditions. The average subjective score for the ODAN-AAD shows a significant improvement over the mixture (56% improvement; paired t test, P < 0.001), demonstrating that the listeners had a stronger preference for the modified audio than for the original mixture. Figure 3c also shows a small but significant difference between the average MOS score with the actual clean sources and that with ODAN separated sources (78% vs. 56% improvement over the mixture). The MOS values using the clean sources show the upper bound of AAD improvement if the speaker separation algorithm was perfect. Therefore, this analysis illustrates the maximum extra gain that can be achieved by improving the accuracy of the speech separation algorithm (14% over the current system).

4   Discussion

We present a framework for AAD that addresses the lack of access to clean speech sources in real-world applications. Our method uses a novel, real-time speaker-independent speech separation algorithm that uses deep-learning methods to separate the speakers from a single channel of audio. Then, the separated sources are compared to the reconstructed spectrogram from the auditory cortical responses of the listener to determine and amplify the attended source. The integration of speaker-independent speech separation in the AAD framework is also a novel contribution. We tested a system on two unseen speakers and showed improved subjective and objective perception of the attended speaker when using the ODAN-AAD framework. A major advantage of our system over previous work [15] is the ability to generalize to unseen speakers, which enables a user to communicate more easily with new people. Because ECoG electrodes reflect the summed activity of thousands of neurons in the proximity of the electrodes [16], the spectral tuning resolution of the electrodes is relatively low [17]. As a result, the reconstruction filters that map the neural responses to the stimulus spectrogram do not have to be trained on specific speakers and can generalize to novel speakers, as we have shown previously [5, 18].

In summary, our proposed speaker-independent AAD system represents a feasible solution for a major obstacle in creating a brain-controlled hearing device, therefore bringing this technology a step closer to reality. Such a device can help hearing impaired listeners more easily communicate in crowded environments and reduce the listening effort for normal hearing subjects, therefore reducing listening fatigue.

Our ongoing research on this problem focuses on: improving the robustness of the system to various noisy, reverberant acoustic conditions; designing the models that require lower computation and power resources for wearable devices; and using noninvasive neural recordings, including scalp EEG with the same or different gender mixtures [6], around the ear EEG electrodes [19], and in-ear EEG recordings [20]. These EEG electrode approaches can further increase the fidelity of the neural recording to improve both the accuracy and speed of attention decoding.

Acknowledgements This work was funded by a grant from the National Institutes of Health, NIDCD-DC014279, National Institute of Mental Health, R21MH114166, and the Pew Charitable Trusts, Pew Biomedical Scholars Program.

This work has been published, and is presented here re-formatted for the BCI Award 2020 [21].

References

1. Carhart R, Tillman TW (1970) Interaction of competing speech signals with hearing losses. Arch Otolaryngol 91(3):273–279
2. Hamacher V, Chalupper J, Eggers J, Fischer E, Kornagel U, Puder H et al (2005) Signal processing in high-end hearing aids: state of the art, challenges, and future trends. EURASIP J Appl Signal Process 2915–2929
3. Chen J, Wang Y, Yoho SE, Wang D, Healy EW (2016) Large-scale training to increase speech intelligibility for hearing-impaired listeners in novel noises. J Acoust Soc Am 139(5):2604–2612
4. Plomp R (1994) Noise, amplification, and compression: considerations of three main issues in hearing aid design. Ear Hear 15(1):2–12
5. Mesgarani N, Chang EF (2012) Selective cortical representation of attended speaker in multi-talker speech perception. Nature 485(7397):233–236
6. O'Sullivan JA, Power AJ, Mesgarani N, Rajaram S, Foxe JJ, Shinn-Cunningham BG et al (2015) Attentional selection in a cocktail party environment can be decoded from single-trial EEG. Cereb Cortex 25(7):1697–1706
7. Wang D, Chen J (2018) Supervised speech separation based on deep learning: an overview. IEEE/ACM Trans Audio Speech Lang Process
8. Luo Y, Chen Z, Mesgarani N (2018) Speaker-independent speech separation with deep attractor network. IEEE/ACM Trans Audio, Speech, Lang Process 26(4):787–796
9. Hershey JR, Chen Z, Le Roux J, Watanabe S (2016) Deep clustering: discriminative embeddings for segmentation and separation. In: IEEE International Conference Acoustics Speech Signal Processing 31–35
10. Kolbæk M, Yu D, Tan Z-H, Jensen J, Kolbaek M, Yu D et al (2017) Multitalker speech separation with utterance-level permutation invariant training of deep recurrent neural networks. IEEE/ACM Trans Audio Speech Lang Process 25(10):1901–1913

11. Strang G, Strang G, Strang G, Strang G (1993) Introduction to linear algebra. Wellesley-Cambridge Press Wellesley, MA
12. Mesgarani N, David SVSV, Fritz JBJB, Shamma SASA (2009) Influence of context and behavior on stimulus reconstruction from neural activity in primary auditory cortex. J Neurophysiol 102(6):3329–3339
13. Brungart DS (2001) Informational and energetic masking effects in the perception of two simultaneous talkers. J Acoust Soc Am 109(3):1101–1109
14. MOS (2006) Vocabulary for performance and quality of service. ITU-T Rec 10
15. O'Sullivan J, Chen Z, Herrero J, McKhann GMGM, Sheth SASA, Mehta ADAD et al (2017) Neural decoding of attentional selection in multi-speaker environments without access to clean sources. J Neural Eng 14(5):56001
16. Ray S, Maunsell HR (2011) Different origins of gamma rhythm and high-gamma activity in macaque visual cortex. PLoS Biol 9 (4)
17. Hullett PW, Hamilton LS, Mesgarani N, Schreiner C, Chang EF (2016) Human superior temporal gyrus organization of spectrotemporal modulation tuning derived from speech stimuli. J Neurosci 36(6):2014–2026
18. Akbari H, Khalighinejad B, Herrero J, Mehta A, Mesgarani N (2018) Reconstructing intelligible speech from the human auditory cortex. BioRxiv 350124
19. Mirkovic B, Debener S, Jaeger M, De Vos M (2015) Decoding the attended speech stream with multi-channel EEG: implications for online, daily-life applications. J Neural Eng 12(4):46007
20. Fiedler L, Wöstmann M, Graversen C, Brandmeyer A, Lunner T, Obleser J (2017) Single-channel in-ear-EEG detects the focus of auditory attention to concurrent tone streams and mixed speech. J Neural Eng 14(3):36020
21. Han C, O'Sullivan J, Luo Y, Herrero J, Mehta AD, Mesgarani N (2019) Speaker-independent auditory attention decoding without access to clean speech sources. Sci Adv 5(5)

A High-Performance Handwriting BCI

Francis R. Willett

Abstract One of the main goals of brain-computer interface (BCI) research is to restore communication to people with people with little or no control of their movements. In this chapter, we interviewed Frank Willett about his work with a BCI to decode handwriting movements. This BCI won first place in the 2020 BCI Research Awards. This interview presents some information about how their BCI system helped a person with paralysis. Their system attained 18 words per minute with very high accuracy. Dr. Willett also talked about challenges for clinical translation and how students and others could get involved.

Keywords Handwriting · Paralysis · Microelectrode arrays · Brain-computer interface (BCI)

1 Introduction

Francis R. Willett, Donald T. Avansino, Leigh Hochberg, Jaimie Henderson and Krishna V. Shenoy submitted their work to the BCI Award 2020 and won 1st place! The team joined forces from Stanford University, the Howard Hughes Medical Institute, Brown University, Harvard Medical School and the Massachusetts General Hospital to decode handwriting with a BCI. We had the chance to talk with Frank about his work. His team also submitted a video with additional information about their 2020 project.[1]

---

[1] https://www.youtube.com/watch?v=wyFj3yl3Aik.

F. R. Willett (✉)
Stanford University, Stanford, USA

Howard Hughes Medical Institute, Chevy Chase, USA

© The Author(s), under exclusive license to Springer Nature Switzerland AG 2021   105
C. Guger et al. (eds.), *Brain-Computer Interface Research*,
SpringerBriefs in Electrical and Computer Engineering,
https://doi.org/10.1007/978-3-030-79287-9_11

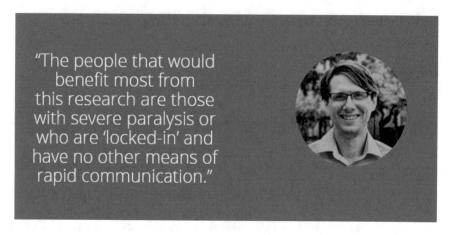

"The people that would benefit most from this research are those with severe paralysis or who are 'locked-in' and have no other means of rapid communication."

This team is not new to the BCI Research Awards. Some people from the same group won first place last year and the preceding year, at the 2018 and 2019 BCI Research Awards. Dr. Willett was nominated in 2014. Dr. Hochberg, a key member of Dr. Willett's team, was an author of a project that was nominated in 2012, and Dr. Hochberg was on the 2012 jury. Like the project they submitted to the 2020 BCI Research Awards, their earlier work that was nominated involved different types of implanted BCIs to help persons with severe difficulty moving. This year's first-place project was:

A High-Performance Handwriting BCI
Francis R. Willett[1,2], Donald T. Avansino[1], Leigh Hochberg[3], Jaimie Henderson[1], Krishna V. Shenoy[1,2]

[1] *Stanford University, USA*

[2] *Howard Hughes Medical Institute, USA*

[3] *Brown University, Harvard Medical School, Massachusetts General Hospital, USA*

2   Interview

Hi Frank, you won 1st place at the International BCI Research Award 2020 competition. Could you briefly describe what this project was about?

Frank: Our project was about building an intracortical BCI to decode handwriting movements in a person with paralysis. This approach can enable someone who is locked-in to type text on a computer by attempting to handwrite it. We demonstrated our BCI in real-time in a person whose hand was paralyzed, and showed that it effectively doubled the prior record for BCI communication rate while achieving high accuracies.

What was your goal?
Frank: Our goal was to demonstrate the feasibility of decoding dexterous behaviors from a person with paralysis, with enough accuracy and speed to significantly improve upon the state of the art in communication BCIs. First, we had to assess whether dexterous handwriting movements could even be decoded at all in someone who has been paralyzed for many years, since to our knowledge no one has done this before. To our surprise, we found that when we asked our participant to attempt to handwrite single letters, they evoked strong and repeatable patterns of neural activity that encoded the pen movement. Our next goal was to see if we could decode complete handwritten sentences, thus allowing someone to communicate a message by attempting to handwrite it.

What technologies did you use?
Frank: We used intracortical microelectrode arrays combined with a variety of computational techniques to achieve high decoding accuracies on this challenging problem. The major challenge we faced was training decoders on data where no overt behavior was available, since our participant's hand was paralyzed. We borrowed techniques from the automatic speech recognition field to solve this problem, using hidden Markov models to infer when our participant wrote each letter in the training data. After completing this inference step, we then used machine learning techniques to train recurrent neural networks to convert the neural activity into the probability of each letter being written at the current time. Finally, we used large-vocabulary, general-purpose language models to autocorrect for occasional decoding errors (Fig. 1).

What kinds of people could benefit from your research?
Frank: The people that would benefit most from this kind of research are those with severe paralysis or who are 'locked-in' and have no other means of rapid communication. One promising thing about this work is that it significantly increases the speed of BCI communication (to approximately 18 words per minute), making it more likely that someone could benefit from this technology even if they have some rudimentary, retained motion.

Do you think your work has future potential for clinical use?
Frank: Yes, we think that the high speed and accuracy that our BCI achieved on general-purpose sentence writing is promising for clinical viability. To our knowledge, this is the fastest communication BCI that is also accurate and general enough to enable the user to write any sentence. It is also entirely self-paced and leaves the eyes free to look anywhere. One remaining challenge, however, is the need to retrain the decoder each day to account for changes in the neural recordings that occur over time. We are currently working on methods to retrain the decoder in an unsupervised way in the background, so that the user does not have to be interrupted for retraining.

Fig. 1 Frank Willett presents his first-place award certificate

This is still an outstanding challenge for BCIs in general, but it is encouraging that many groups are beginning to tackle this problem. I think it's likely that a combination of algorithmic innovation on this front, combined with improvements to device stability, will continue to improve the robustness of intracortical BCIs.

What was it like to win the BCI Award 2020?
Frank: It's great to be recognized as doing useful and interesting research, and I'm thankful that the field has the BCI awards as a place to highlight and be inspired by the latest BCI developments.

How can students and other researchers get involved in such research?
Frank: Our work is highly interdisciplinary, intersecting with hardware and device design, computational techniques from computer science and machine learning, and

basic neuroscience. Students and researchers can contribute by working on the next generation of electrode technology, designing new algorithms for decoding neural activity, and expanding our knowledge of the brain circuits that underly the control of movement. In the intracortical BCI field, more labs are now getting involved in clinical trials, as we have done here. In addition, new companies are joining the field (e.g. Neuralink, Paradromics). The future looks bright for BCIs!

A Neuromorphic Brain-Computer Interface for Real-Time Detection of a New Biomarker for Epilepsy Surgery

Karla Burelo

Abstract The annual BCI Research Awards highlight each year's best projects involving BCIs. This year, one of the submissions that was nominated for an award introduced a new approach to help surgeons identify areas that cause seizures. This new work could make epilepsy surgeries more effective while helping us understand what causes epilepsy and why medications to help people with epilepsy are not always effective. Future work could help other types of patients and even identify patterns outside of the brain. This chapter presents an interview with Karla Burelo, the lead author of this project.

Keywords Epilepsy · EEG · ECoG · High-frequency oscillations (HFOs) · Brain-computer interface (BCI)

1 Introduction

Some patients with epilepsy can use medications to reduce their seizures. However, about a quarter of these patients have drug-resistant epilepsy or DRE [1–2]. For them, surgery may be the best option to reduce or eliminate seizures by removing the brain areas that trigger seizures. Karla Burelo and her team were nominated for a BCI Research Award for their project, which was:

Karla Burelo[1,2], Mohammadali Sharifshazileh[1,2], Johannes Sarnthein[2], and Giacomo Indiveri[1]

A neuromorphic brain computer interface for real-time detection of a new biomarker for epilepsy surgery

1 University of Zurich and ETH Zurich, Institute of Neuroinformatics, Switzerland
2 University Hospital and University of Zurich, Switzerland

K. Burelo (✉)
Institute of Neuroinformatics, University of Zurich and ETH Zurich, Zurich, Switzerland

University Hospital and University of Zurich, Zurich, Switzerland

© The Author(s), under exclusive license to Springer Nature Switzerland AG 2021   111
C. Guger et al. (eds.), *Brain-Computer Interface Research*,
SpringerBriefs in Electrical and Computer Engineering,
https://doi.org/10.1007/978-3-030-79287-9_12

Fig. 1 The members of the HFO group: Ece Boran, Mohammadali Sharifshazileh, Tommaso Fedele, Karla Burelo, Giacomo Indiveri and Johannes Sarnthein. The HFO group is a synergy between the Institute of Neuroinformatics, UZH and ETH, and the Clinic for Neurosurgery, USZ

We interviewed the lead author, Karla Burelo, to learn more about how she developed a new approach to improve epilepsy surgery based on High Frequency Oscillations (HFOs) with her team. The interview shows how the team had to consider hardware, software, and algorithms. They also addressed both invasive (ECoG) and non-invasive (EEG) ways to measure brain activity, because surgeons often use EEG to identify regions of interest and then use ECoG for more precise detail. Figure 1 shows Karla and her team, who also contributed a video about their project.[1]

2 Interview

You submitted your project about a neuromorphic brain-computer interface for real-time detection of a new biomarker for epilepsy surgery to the BCI Award

---

[1] https://www.youtube.com/watch?v=Pw83Mrza_rg.

2020, and you were nominated with this great project. Can you describe what this project is about?

Karla: Sure. Our project tried to combine neuromorphic engineering [3, 4, 5, 6, 7, 8] and a new way to record the signals that provides a better signal to noise ratio. We want to provide a new device that can be used for long term monitoring or during surgery. That is the big scope of the project. Here, we focus on explaining all the hardware engineering we've been doing to provide the device, and the software techniques that we use for a neural network that can solve the task of finding the biomarkers.

How did you distribute the work in your team?

Karla: Mohammad is mostly in charge of designing all the hardware. He designed the filters and amplifiers that are connected with the neuromorphic chip. He didn't design the chip—a group in our lab designed it—but Mohammad designed all the front end, including the amplifiers, filters, and signal-to-spike conversion, which is very important [9]. On my side, I looked at how to build a spiking neural network that can solve the task of finding these patterns, but also can be mapped into the hardware we have [10]. We were not going to have them both at the same time, so these tasks were in parallel. I was working on the spiking network trying to constrain to what the hardware would do (Fig. 2). The idea to come up with this project was from Giacomo and Johannes. Johannes does the intracranial recordings together with the doctors and Giacomo has developed the ideas for neuromorphic engineering.

However, a lot of people are involved in the current state of the project and I would like to acknowledge the Neuromorphic Cognitive Group for developing the software to use the chip, developing the circuits and building blocks in the rest of the chip, for discussing together, and for the general support.

Maybe you can explain the different tools that you used for this project?

Karla: For the hardware design, we used Cadence®. We also used Python™ for the low-level software framework. We collaborated with a company called SynSense in Switzerland. They also have the rights to the chip with the spiking neural network. There's a collaboration to have the software tools to interface with the hardware, which is called SAMNA. This is the software that you can use to talk to the chip if you want to send some spikes or just test the spiking neural network. For the software, we used Python. We also used Brian2 [11], which is a simulator for spiking neural networks. It's used to solve differential equations for a spiking neural network. There's another toolbox that was designed here at the Institute of Neuroinformatics that is called Teili [12]. This is a toolbox on top of Brian2 to solve some minor issues with the connections of the neurons. It also wraps around the equations that we have that describe the hardware. It's a library to use exactly the equations that we have in hardware and thus have a better simulation of our chips.

Let's take one step backwards. Can you provide a high-level description of your project?

Karla: Yes, we focus on people with severe epilepsy. Some people cannot benefit from drugs, so they need surgery to be cured of epilepsy. In some of these people, it may be an option that a surgeon removes the area of the brain that causes the seizures. But finding the epileptogenic area (that is, the area that causes the seizures) is challenging. There are many studies to try to find these areas. Once the surgeon has removed what he thought was the epileptogenic zone, one continues to check after six, twelve, or 24 months whether the patients have seizure nevertheless [13, 14]. The success rate of this surgery is only about 60%—meaning that the patient never has another seizure. That's why we would like to find something to increase this success rate.

It has been proposed that high-frequency oscillations or HFOs are good biomarkers for the epileptogenic zone that need to be removed [15]. Nowadays, HFOs are not widely used. Rather, there are MRIs and sometimes recordings with EEG electrodes implanted inside the brain. To further analyze and find these HFOs and indicate epileptogenic areas, we use these intracranial EEG signals. These signals are recorded while the patient is in a hospital for several days and/or during surgery [16, 17–18, 15]. When the patient is in surgery, surgeons can place electrodes on the brain to see which areas are important that should not be removed. Since that is already part of surgical procedure, we want a device that can directly interact with these electrodes. We could then detect these HFOs in real-time and detect areas with most of these patterns, and then guide the surgery [9, 19].

If you want to introduce such a device into the surgery room, you have to consider some constraints. For example, you need something compact, because you don't want more bulky devices. It also has to be battery powered, since we are trying to detect very small signals, and we don't want to create noise from other electronics that could affect the classification task.

These HFOs are tricky and tiny. Are you able to pick them up with scalp EEG?

Karla: That is actually a new thing. Several researchers have found meaningful HFOs in the scalp EEG, often in children, also in Zurich [20, 21]. With a long-term recording over weeks or months we might be able to tell the severity of the epilepsy, whether some treatment actually helped, or whether the patient can even be relieved from the anti-epileptic drugs that have many side-effects. We analyzed some scalp EEG recordings and found similar HFOs like standard HFO detectors do. But these detectors need a lot of computer power and a battery would have to be recharged very often. So long-term recordings is certainly where we want to go in the future, because there the extremely low power consumption of our neuromorphic device is a real advantage.

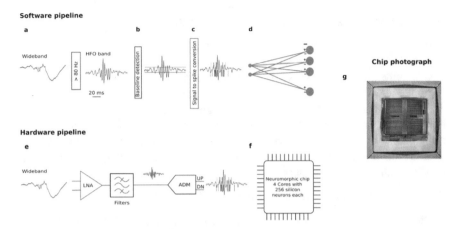

Fig. 2 A neuromorphic system that combines a headstage, band-pass filtering stages, a signal-to-spike conversion circuit and a multi-core SNN architecture for recording, processing, and detecting clinically relevant HFO in iEEG from TLE (temporal lobe epilepsy) patients

What kind of people could benefit from your research? Is it just for people with epilepsy?

Karla: Yes, we are currently focused on people with epilepsy who are going into surgery. That's the approach we have at the beginning of this effort, especially because the signal to noise ratio is most favorable when we record directly from inside the brain. But there are also other options. The spiking neural network that I designed is programmable. The analog head-stage works for anything; you just analyze the data with any spiking neural network that you want. Ideally, I could design a spiking neural network to detect patterns in the heart or any muscle. So, we could really expand this neuromorphic analysis technique to other types of data recorded from patients with other diseases. As soon as we have a signal that we could interface with a device, it's a matter of finding a spiking neural network that can accomplish a specific task to help that patient.

What's it like to work at ETH and the University of Zurich?

Karla: It's very nice. I really like it. I especially like that the institute is very interdisciplinary. We have people from disciplines such as psychology, biology, mathematics, and electrical engineering. I'm personally a chemical engineer. It's a very nice environment. The university hospital is also excellent. I have been able to go to surgery, just watching from the corner. I could observe how everything is done, how many machines are involved, and how everything is very precise. It's a nice combination to have this interdisciplinary environment, including close contact with patients in the hospital.

How could students get involved in your research?

Karla: We have a student who is helping with the testing of the chip. Mohammad designed a chip, and a lot of tests have to be done. This gives Mohammad time to design more chips. There are some projects with designing other blocks using other technologies. We have a girl doing chip design, and we have a guy who is testing the resulting chips. On the software side, we have a student who will try to analyze how neurons in the chips work, in terms of how they work as filters and how we can combine them to increase accuracy. It's really nice that we have these Teili and Brian2 tools to simulate what the hardware does. So, there are plenty of possibilities to do a small project or masters' thesis.

What was it like to be nominated for the BCI Award?

Karla: We were very excited. We were very happy. We have a colleague who really likes to record video, and it was fun to develop the video that we submitted for the award, even if we weren't nominated. We enjoyed making the video because we sat down and tried to sketch everything we did to decide how to present it within a two-minute video. That brought us a lot of joy. Sometimes, you don't look back, you just produce work. But we could look back and say: "This is what we have done." When we got nominated, we said: "Wow, we won something!".

References

1. Dalic L, Cook MJ (2016) Managing drug-resistant epilepsy: challenges and solutions. Neuropsychiatr Dis Treat 12:2605
2. Kwan P, Schachter SC, Brodie MJ (2011) Drug-resistant epilepsy. N Engl J Med 365(10):919–926
3. Mead C (2020) How we created neuromorphic engineering. Nat Electron 3:434–435
4. Chicca E, Stefanini F, Bartolozzi C, Indiveri G (2014) Neuromorphic electronic circuits for building autonomous cognitive systems. Proc IEEE 102:1367–1388
5. Corradi F, Indiveri G (2015) A neuromorphic event-based neural recording system for smart brain-machine-interfaces. IEEE Trans Biomed Circuits Syst 9:699–709
6. Indiveri G, Liu SC (2015) Memory and information processing in neuromorphic systems. Proc IEEE 103:1379–1397
7. Qiao N, Mostafa H, Corradi F, Osswald M, Stefanini F, Sumislawska D, Indiveri G (2015) A reconfigurable on-line learning spiking neuromorphic processor comprising 256 neurons and 128K synapses. Front Neurosci 9:141
8. Rubino A, Livanelioglu C, Qiao N, Payvand M, Indiveri G (2021) Ultra-low-power FDSOI neural circuits for extreme-edge neuromorphic intelligence. IEEE Trans Circ Syst I Regul Pap 68:45–56
9. Sharifshazileh M, Burelo K, Fedele T, Sarnthein J, Indiveri G (2019) A neuromorphic device for detecting high-frequency oscillations in human iEEG. In: 26th IEEE international conference on electronics, circuits and systems (ICECS). IEEE, pp 69–72
10. Burelo K, Sharifshazileh M, Krayenbühl N, Ramantani G, Indiveri G, Sarnthein J (2021) A spiking neural network (SNN) for detecting high frequency oscillations (HFOs) in the intraoperative ECoG. Sci Rep 11:6719
11. Goodman D, Brette R (2008) Brian: a simulator for spiking neural networks in python. Front Neuroinformatics 2 (2008). http://doi.org/10.3389/neuro.01.026.2009

12. Milde M et al (2018) teili: a toolbox for building and testing neural algorithms and computational primitives using spiking neurons. Unreleased software, Institute of Neuroinformatics, University of Zurich and ETH Zurich
13. Ramantani G (2019) Epilepsy surgery in early life: the earlier, the better. World Neurosurg 131:285–286
14. Ryvlin P, Rheims S (2016) Predicting epilepsy surgery outcome. Curr Opin Neurol 29:182–188
15. van't Klooster MA, van Klink NE, Leijten FS, Zelmann R, Gebbink TA, Gosselaar PH, Braun KP, Huiskamp GJ, Zijlmans M (2015) Residual fast ripples in the intraoperative corticogram predict epilepsy surgery outcome. Neurology 85:120–128
16. Boran E, Ramantani G, Krayenbuhl N, Schreiber M, Konig K, Fedele T, Sarnthein J (2019) High-density ECoG improves the detection of high frequency oscillations that predict seizure outcome. Clin Neurophysiol 130:1882–1888
17. Fedele T, Burnos S, Boran E, Krayenbühl N, Hilfiker P, Grunwald T, Sarnthein J (2017) Resection of high frequency oscillations predicts seizure outcome in the individual patient. Sci Rep 7:13836
18. Fedele T, van't Klooster M, Burnos S, Zweiphenning W, van Klink N, Leijten F, Zijlmans M, Sarnthein J (2016) Automatic detection of high frequency oscillations during epilepsy surgery predicts seizure outcome. Clin Neurophysiol 127:3066–3074
19. Sharifshazileh M, Burelo K, Sarnthein J, Indiveri G (2021) An electronic neuromorphic system for real-time detection of high frequency oscillations (HFOs) in intracranial EEG. Nat Commun 12(1):1–14
20. Boran E, Sarnthein J, Krayenbuhl N, Ramantani G, Fedele T (2019) High-frequency oscillations in scalp EEG mirror seizure frequency in pediatric focal epilepsy. Sci Rep 9:16560
21. Cserpan D, Boran E, Lo Biundo SP, Rosch R, Sarnthein J, Ramantani G (2021) Scalp high-frequency oscillation rates are higher in younger children. Brain Commun 3(2)
22. Kalilani L, Sun X, Pelgrims B, Noack-Rink M, Villanueva V (2018) The epidemiology of drug-resistant epilepsy: a systematic review and meta-analysis. Epilepsia 59(12):2179–2193
23. Fedele T, Ramantani G, Burnos S, Hilfiker P, Curio G, Grunwald T, Krayenbuhl N, Sarnthein J (2017) Prediction of seizure outcome improved by fast ripples detected in low-noise intraoperative corticogram. Clin Neurophysiol 128:1220–1226
24. Fedele T, Ramantani G, Sarnthein J (2019) High frequency oscillations as markers of epileptogenic tissue—end of the party? Clin Neurophysiol 130:624–626

Recent Advances in Brain-Computer Interface Research: A Summary of the 2020 BCI Award

Christoph Guger, Brendan Z. Allison, and Aysegul Gunduz

Abstract We began this book with an introductory chapter so readers could understand more about BCIs and the annual procedures we follow to develop the awards and these books. The subsequent chapters of this book each presented a BCI project that was nominated for a BCI Research Award, with seven project summaries and four interviews. In the concluding chapters of the previous several books in this series, we have presented the first, second, and third place winners. This year, we still present the three winners—but we had no third-place winner. For the first time, we had a tie for second place this year. We also hosted our first Awards Ceremony online due to COVID. This chapter also discusses our plans for next year's BCI Research Award and future directions.

Keywords Brain-computer interface · EEG · ECoG · BCI Research Awards · BCI Foundation

1 The 2020 Awards Ceremony

The Awards Ceremony was part of the virtual IEEE Systems, Man, and Cybernetics conference in October 2020. As with prior years, the nominees were told that the first, second, and third place winning teams would earn $3000, $2000, and $1000, respectively—in addition to the prestige of winning. The prizes were generously donated by the BCI Society, IEEE Brain, Cortec GmbH and the main sponsor g.tec medical engineering GmbH. The BCI Society is a non-profit organization that organizes the

C. Guger (✉)
g.Tec Medical Engineering GmbH, Schiedlberg, Austria
e-mail: guger@gtec.at

B. Z. Allison
Cognitive Science Department, University of California at San Diego, San Diego, USA

A. Gunduz
Biomedical Engineering, University of Florida, Gainesville, USA

BCI Meeting series (bcisociety.org). The mission of IEEE Brain is to facilitate cross-disciplinary collaboration and coordination to advance research, standardization and development of technologies in neuroscience to help improve the human condition (brain.ieee.org). Cortec GmbH is a company from Germany that manufactures high-quality ECoG technology and implants. g.tec medical engineering designs and manufactures high-quality equipment and software for BCIs and other applications. The BCI Award Foundation organized the 2020 BCI Research Award.

2 The 2020 Winners

The winners of the BCI Award 2020 were announced in October 2020 at the IEEE SMC conference[1] during the Award ceremony. The head of the jury was Aysegul Gunduz. The other jury members were Sergey D. Stavisky (winner 2019), Adriane Randolph, Steve Meng, Jörn Rickert, Fabien Lotte and Yannick Roy. The jury selected twelve nominees and three winners, which were:

First Place Winner:

A High-Performance Handwriting BCI

Francis R. Willett[1,2], Donald T. Avansino[1], Leigh Hochberg[3], Jaimie Henderson[1], Krishna V. Shenoy[1,2]

[1] Stanford University, USA

[2] Howard Hughes Medical Institute, USA

[3] Brown University, Harvard Medical School, Massachusetts General Hospital, USA

Second Place Winner:

High-dimensional (8D) Control of Complex Effectors such as an Exoskeleton by a Tetraplegic Subject Using Chronic ECoG Recordings Using Stable and Robust Over Time Adaptive Direct Neural Decoder

Alexandre Moly[1], Thomas Costecalde[1], Félix Martel[1], Antoine Lassauce[1], Serpil Karakas[1], Gael Reganha[1], Alexandre Verney[2], Benoit Milville[2], Guillaume Charvet[1], Stéphan Chabardes[3], Alim Louis Benabid[1], Tetiana Aksenova[1]

[1] CEA, LETI, CLINATEC, University Grenoble Alpes, MINATEC, France

[2] CEA, LIST, DIASI, SRI, Gif-sur-Yvette, France

[3] Centre Hospitalier Universitaire Grenoble Alpes, France

Second Second Place Winner:

---

[1] http://smc2020.org/.

A Brain-Spine Interface Complements Deep-Brain Stimulation to Both Alleviate Gait and Balance Deficits and Increase Alertness in a Primate Model of Parkinson's Disease

Tomislav Milekovic[1,2,3,4], Flavio Raschellà[1,2,3,5], Matthew G. Perich2, Eduardo Martin Moraud[1,2,3,6], Shiqi Sun[1,2,3,7], Giuseppe Schiavone[8], Yang Jianzhong[9,10], Andrea Galvez[1,2,3,4], Christopher Hitz[1], Alessio Salomon[1], Jimmy Ravier[1,2,3], David Borton[1,11], Jean Laurens[1,12], Isabelle Vollenweider[1], Simon Borgognon[1,2,3], Jean-Baptiste Mignardot[1], Wai Kin D Ko[9,10], Cheng YunLong[9,10], Li Hao[9,10], Peng Hao[9,10], Laurent Petit[13,14], Qin Li[9,10], Marco Capogrosso[1], Tim Denison[15], Stéphanie P. Lacour[8], Silvestro Micera[5,16], Chuan Qin[10], Jocelyne Bloch[1,2,3,6], Erwan Bezard[9,10, 13,14], Grégoire Courtine[1,2,3,6]

[1] Center for Neuroprosthetics (CNP) and Brain Mind Institute, School of Life Sciences, Swiss Federal Institute of Technology (EPFL), Switzerland

[2] Department of Clinical Neuroscience, Lausanne University Hospital (CHUV) and University of Lausanne (UNIL), Switzerland

[3] Defitech Center for Interventional Neurotherapies (NeuroRestore), CHUV/UNIL/EPFL, Switzerland

[4] Department of Fundamental Neuroscience, Faculty of Medicine, University of Geneva, Switzerland

[5] CNP and Institute of Bioengineering, School of Engineering, EPFL, Switzerland

[6] Department of Neurosurgery, CHUV, Switzerland

[7] Beijing Engineering Research Center for Intelligent Rehabilitation, College of Engineering, Peking University, People's Republic of China

[8] CNP, Institute of Microengineering and Institute of Bioengineering, School of Engineering, EPFL, Switzerland

[9] Motac Neuroscience, UK

[10] Institute of Laboratory Animal Sciences, China Academy of Medical Sciences, People's Republic of China

[11] Carney Institute for Brain Science, School of Engineering, Brown University, USA

[12] Department of Neuroscience, Baylor College of Medicine, USA

[13] Université de Bordeaux, Institut des Maladies Neurodégénératives (IMN), UMR 5293, France

[14] CNRS, IMN, UMR 5293, France

[15] Oxford University, UK

[16] The BioRobotics Institute, Scuola Superiore Sant'Anna, Italy

Aysegul Gunduz: "The three winners are doing cutting edge research and are showing the whole complexity that is involved in BCI development".

This year, we had a second place winner—our first tie for second place! Congratulations to all three of the winning teams, and thanks to them (and other nominees) for their chapters in this book. We also thank the sponsors, jury, and everyone who submitted their BCI project for an award.

The three winning projects gave a key-note talk at the g.tec BCI & Neurotechnology Spring School 2021 in front of 4025 people and answered many questions from the attendees.

3 Conclusion with Past and Future Directions

This is the ten-year anniversary of the BCI Research Awards, and the tenth book devoted to these awards. Hence, this section briefly reviews the ten years of the awards and the books, and concludes with future directions for these awards. The first book based on the BCI Research Awards was published through InTech Open, and introduced the 2010 BCI Research Awards [1]. The introductory chapter from that book is available for free online.[2] We then began this series of books titled "A State-of-the-Art Summary" with the review of the second BCI Research Awards from 2011 [2]. So, every book about these awards except the first book was part of the same series with Springer Publishing. The book for the 2021 BCI Research Awards will be with the same publisher.

The first BCI Research Award occurred in 2010. The first Gala Awards Ceremony was held at the Fifth International BCI Meeting at Asilomar, California—the same location where three later BCI Awards ceremonies were held (in 2012, 2015, and 2018). Some components of the award have changed, including:

- We had only one winner each year for the first awards through 2013. In 2014, we introduced the format we use today with first, second, and third place winners.
- Each annual award had 10 nominees until 2016. Since the 2016 awards, we chose twelve nominees each year.
- The cash prizes increased over the years, and some of the other prizes have changed. For example, a few years ago, we began awarding a Pfurtscheller bread knife.
- In 2017, we started the non-profit BCI Award Foundation to manage the BCI Research Awards. The first several BCI Research Awards were administered by g.tec medical engineering GmbH.
- We added interviews in the book that presents the 2018 awards as part of the discussion chapter [3]. Last year and this year, we developed interviews into chapters.

---

[2] https://www.intechopen.com/books/recent-advances-in-brain-computer-interface-systems/state-of-the-art-in-bci-research-bci-award-2010.

- The first several books included discussion of trends from the concluding chapter. We decided to instead present more commentary in the concluding chapters.
- The sponsors changed over the years.
- We had our first Awards Ceremony online to present the 2020 BCI Research Awards.

However, many central components of the BCI Research Award have not changed much since the awards began ten years ago. Key elements have not changed, including the submission rules, judging criteria, jury selection procedure, Awards Ceremony to announce the winners, and publication of an annual book reviewing the Awards. Drs. Guger and Allison have emceed all of the in-person Awards Ceremonies, announcing over 100 nominees and a dozen winners over the ten years of the Awards.

Most importantly, the BCI Research Awards and books have never changed their altruistic goals. We wanted to identify and publicly recognize the best annual BCI projects, inform readers about a research field that we love, draw attention to BCI research and development, and encourage the best future BCI projects. We wanted to select nominees and winners from anywhere in the world, without regard for the type of equipment used, using a jury with the best BCI experts worldwide. The juries have always had 5–8 people and included a range of experts who can judge different aspects of BCI projects and understand medical, scientific, and technical facets of BCI projects.

As of this writing (April 2021), we have selected the jury for the 2021 BCI Research Awards. Figure 1 shows that the 2021 jury has seven BCI practitioners,

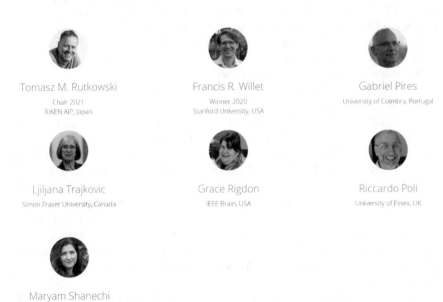

Fig. 1 The jury for the 2021 BCI Research Awards

Fig. 2 Sponsors for the 2021 BCI Research Awards

most of whom are experienced with BCI Research Awards. Like most years, our jury includes a member of the team that won first place the preceding year. This time, that team member is Dr. Francis R. Willett, who has a chapter in this book with an interview about his first-place project in 2020. The chair, Prof. Tomasz "Tomek" M. Rutkowski, won first place in the 2014 BCI Research Award for an airborne ultrasonic "display." Prof. Maryam Shanechi and her team won third place in the 2019 BCI Research Awards, and she led projects that were nominated for BCI Research Awards in 2013 and 2014.

Prof. Ljiljana Trakjovic organized a "BMI Workshop" with Dr. Guger at the 2020 IEEE SMC conference where we held the 2020 BCI Research Awards Ceremony, and Prof. Grace Rigdon was a juror of a BR4IN.IO hackathon. Overall, the jury for 2021 (like the 2020 jury and nominees) reflects a good mix of people who are experienced with the BCI Research Awards and new people.

We also announced the key dates for the 2021 awards:

Submission deadline: August 1

Announcement of nominees: September 5

Oral presentations from nominees: October 18

The BCI Award Ceremony: October 19.

The BCI Award Ceremony will occur at an online IEEE Systems, Man, and Cybernetics conference, like the 2020 Award Ceremony. This year, this conference[3] will be hosted virtually from Melbourne, Australia from October 17–20. The conference will also feature the 11th BMI Workshop and numerous other activities involving BCI/BMI.

Figure 2 shows that the 2021 awards will have more sponsors than ever before. Like the 2020 awards, the sponsors include g.tec medical engineering, CorTec, and IEEE Brain. We have four new sponsors. Two of them are companies (Intheon and AIP). Riken is a famous brain research institute in Japan, and NeurotechX is a network of organizations at universities and research centers devoted to furthering research in neurotechnology.

This increase in sponsors reflects growth of the BCI Research Awards, but is catalyzed by the broader growth in overall BCI R&D. New groups of patients are

---

[3] http://ieeesmc2021.org.

benefiting from BCIs, and consumer BCIs are steadily becoming more common. These awards and books have helped hundreds of thousands of readers to learn more about BCIs. Many people who read these books have gone on to study BCIs, teach classes, develop BCIs, and even earn a nomination or win in a BCI Research Award. BCI research needs people from many disciplines and backgrounds, and our readers might enjoy and contribute to BCIs in different ways. We hope you liked this chapter and the rest of our book, and look forward to more of them.

References

1. Guger C, Bin G, Gao X, Guo J, Hong B, Liu T (2011) State-of-the-art in BCI research: BCI Award 2010. INTECH Open Access Publisher
2. Guger C, Allison BZ, Edlinger G (2013) Brain-computer interface research: a state-of-the-art summary. Springer, Berlin, Heidelberg
3. Guger C, Allison BZ, Miller K (2020) Highlights and interviews with winners. In: Brain–computer interface research. Springer, Cham, pp 107–121

Printed in the United States
by Baker & Taylor Publisher Services